REVISE AQA GCSE (9–1)
Mathematics

REVISION WORKBOOK

Foundation

Series Consultant: Harry Smith

Author: Glyn Payne

Also available to support your revision:

Revise GCSE Study Skills Guide   9781447967071

The **Revise GCSE Study Skills Guide** is full of tried-and-trusted hints and tips for how to learn more effectively. It gives you techniques to help you achieve your best – throughout your GCSE studies and beyond!

Revise GCSE Revision Planner   9781447967828

The **Revise GCSE Revision Planner** helps you to plan and organise your time, step-by-step, throughout your GCSE revision. Use this book and wall chart to mastermind your revision.

> **For the full range of Pearson revision titles across KS2, KS3, GCSE, Functional Skills, AS/A Level and BTEC visit:**
> www.pearsonschools.co.uk/revise

Contents

A small bit of small print
AQA publishes Sample Assessment Material and the
Specification on its website. This is the official content and
this book should be used in conjunction with it. The questions
in 'Now try this' have been written to help you practise every
topic in the book. Remember: the real exam questions may
not look like this.

Place value

1 (a) Write the number nine thousand, three hundred and fifty-one in figures.

Guided

9 ...~...... **(1 mark)**

(b) Write the number 4196 in words.

Four thousand, one hundred and .... **(1 mark)**

(c) Write down the value of the 5 in the number 95872

5 .... **(1 mark)**

2 Write down the number twelve thousand and sixty in a place value table.

Guided

....	....	Hundreds	....	Units
....	....	....	6	0

(2 marks)

3 Write these numbers in order, smallest first.

(a) 165, 146, 127, 49, 169

.... **(1 mark)**

(b) 7429, 7249, 7942, 7924, 7028

.... **(1 mark)**

4 Peter has these number cards.   ☐ 3   ☐ 4   ☐ 7

(a) Use all three cards to write down all the possible three-digit numbers.

....

.... **(2 marks)**

(b) Write the three-digit numbers from part (a) in order, starting with the smallest.

.... **(1 mark)**

5 Here are four number cards.   ☐ 6   ☐ 4   ☐ 8   ☐ 1

Which of these is the largest even number that you can make with these cards?
Circle your answer.

8146   8641   8614   8416   **(1 mark)**

PROBLEM SOLVED!

6 Anton is buying supplies for a charity event.
A pack of 50 paper cups costs £1.89
A pack of 10 paper plates costs 49p
Anton has £15 to spend.
Anton buys 250 paper cups and spends the
rest on paper plates.
How many paper plates can he buy?

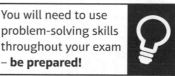

You will need to use
problem-solving skills
throughout your exam
– **be prepared!**

.... **(1 mark)**

Negative numbers

1 Write these numbers in order.

Guided

    6   −11̶   −4   0   4

> Start with the lowest number.

    −11   ....   **(1 mark)**

2 Work out

> − − = +

(a) $-9 + 7 = $ .... **(1 mark)**

(b) $-7 - -4 = $ .... = ... **(1 mark)**

(c) $-6 - 4 = $ .... **(1 mark)**

(d) $-10 - + 6 = $ .... **(1 mark)**

3 On a certain day in Moscow the temperature at 12 noon is 7 °C but by 6 pm it has dropped by 9 °C. By 9 pm it has dropped a further 5 °C and by 12 midnight it has dropped a further 8 °C. Work out the temperature at

(a) 6 pm ....°C **(1 mark)** (b) 9 pm ....°C **(1 mark)** (c) 12 midnight ....°C **(1 mark)**

Guided

(d) What was the overall drop in temperature from 12 noon to 12 midnight?

Temperature at 12 noon = 7 °C

Temperature at 12 midnight = ...

Drop in temperature = 7 − .... = ...°C   **(1 mark)**

4 The table gives information about the highest and lowest temperatures in five cities during one year.

	London	New York	Moscow	New Delhi	Lisbon
Highest temperature (°C)	30	28	25	40	33
Lowest temperature (°C)	−8	−10	−15	−7	−3

(a) What is the difference between the highest temperature in New Delhi and the lowest temperature in New York?

> Write down 'highest temperature of New Delhi – lowest temperature of New York'.

    ....°C   **(1 mark)**

(b) Which city recorded the biggest difference between the highest and lowest temperature?

    ....   **(1 mark)**

Viktor works out that the temperature halfway between the lowest temperature in Lisbon and the lowest temperature in Moscow is –10 °C.

(c) Is Viktor correct? Give a reason for your answer.

    ....   **(1 mark)**

5 Work out

(a) $-7 \times 2 = $ .... **(1 mark)**

(b) $63 \div -9 = $ .... **(1 mark)**

(c) $-6 \times -4 = $ .... **(1 mark)**

(d) $-42 \div -6 = $ .... **(1 mark)**

Rounding numbers

1 Round

(a) 26 723 to the nearest thousand

(b) 6453 to the nearest hundred

(c) 87 536 to the nearest ten.

27 000 **(1 mark)**

6........................ **(1 mark)**

87 5................... **(1 mark)**

2 Round 8.63559 correct to

(a) 1 decimal place

(b) 3 decimal places

(c) 3 significant figures.

.... **(1 mark)**

.... **(1 mark)**

.... **(1 mark)**

3 Round 0.0034672 correct to

(a) 1 significant figure

(b) 2 significant figures

(c) 5 decimal places.

0.003 **(1 mark)**

.... **(1 mark)**

.... **(1 mark)**

4 Round 38 652 correct to

(a) 1 significant figure

(b) 2 significant figures

(c) 3 significant figures.

40 000 **(1 mark)**

.... **(1 mark)**

.... **(1 mark)**

> You need to include enough zeros to show the correct place value.

5 In her science class, Anjali measured the mass of some objects made from different types of materials. Here are her results.

Material	Wood	Plastic	Metal	Rubber
Mass m (g)	20.356	265.800	168.240	127.500

Write down the mass of the

(a) wood to 3 significant figures

(b) metal to 2 significant figures.

.... **(1 mark)**

.... **(1 mark)**

(c) Which of these is the mass of plastic to 1 significant figure?
Circle your answer. **(1 mark)**

265.8   300   200   270

6 Jason is weighing some objects on an electronic scale.   0.02346

He writes the answer as 0.023 g to 3 significant figures.
Is he correct? Give a reason for your answer.

> Which number is the first significant figure?

....

.... **(1 mark)**

Adding and subtracting

1 Work out

(a) 842 + 158 + 23

```
  842
  158
+  23
.......3
```

(1 mark)

(b) 741 − 164

```
      13
  6⁷7̶ 4̶ 1¹
−   1 6 4
  ........7
```

(1 mark)

2 Work out

(a) 7263 + 915

(b) 7629 − 7452

.... **(1 mark)**

.... **(1 mark)**

3 There are 52 children on the pirate ship at a fairground. When the pirate ship stops, 39 children get off and 28 children get on. How many children are now on the pirate ship?

> In this case 'get off' means subtract and 'get on' means add.

Circle your answer.

51   63   55   41   **(1 mark)**

4 Kevin buys some items from a shop. He buys a box of chocolates costing £1.65 and three rolls of wrapping paper costing £0.85 each. He gives the cashier a £10 note. How much change should he receive?

> Convert the pounds into pence.

165 + 85 + 85 + 85 = ...

1000 − .... = ...   **(3 marks)**

5 Part of a receipt is missing. David pays £5 and receives 50p change. David works out that the coffee cost £2.49

Slice of cake	95p
Mug of tea	£1.49
Cup of coffee	

Is he correct? Give a reason for your answer.

....

....

.... **(3 marks)**

4

Multiplying and dividing

1 Work out

 Guided

(a) 83×23

```
      83
  ×   23
  ........
  ........0
  ........   (1 mark)
```

Work out 83×3

Work out 83×20 by writing down a 0 and then working out 83×2

(b) $972 \div 4$

```
    ..........
  4)9¹72
```

$4 \times 2 = 8$ so 4 divides into 9 twice with remainder 1

(1 mark)

2 Tins of biscuits come in three sizes. There are 28 biscuits in the small size and four times as many in the medium size. In the large size there are seven times as many as in the small size. How many more biscuits are in the large size tin than the medium size tin?

Circle your answer.

196   84   122   112   **(1 mark)**

3 A shop sold 42 boxes of flowers. Each box contained 18 flowers.
Work out the total number of flowers sold.

.... **(2 marks)**

4 Work out

Guided

(a) 962×45

```
     962
  ×   45
  ........
  ........0
  ........   (2 marks)
```

Work out 962×5

Work out 962×40 by writing down a 0 and then working out 962×4

(b) $442 \div 13$

```
        3...
  13)442
```

(2 marks)

5 Dylan packs tomato tins into boxes. Each box holds 36 tomato tins.
How many boxes will he need to pack for

(a) 180 tins?

(b) 324 tins?

....

.... **(2 marks)**   .... **(2 marks)**

6 Sam bought five boxes of chocolates. Each box contained 25 chocolates.
Sam ate 30 chocolates himself. He then shared the remaining chocolates
equally between himself and his four friends.

(a) How many chocolates did Sam buy?

(b) How many chocolates did each of Sam's friends receive?

.... **(1 mark)**   .... **(2 marks)**

Decimals and place value

1 (a) Write down the value of the 7 in 9.74

.... **(1 mark)**

> Remember the first number after the decimal point is a tenth and then a hundredth and so on.

(b) Write down the value of the 8 in 0.684

.... **(1 mark)**

(c) Write down the value of the 4 in 0.704

.... **(1 mark)**

2 Write these numbers in order, smallest first.

3.2   6.4   6.2   12.8   1.4

.... **(1 mark)**

Guided

3 Write these numbers in order, smallest first.

0.61   0.611   0.613   0.6   0.05

0.610   0.611   0.613   0.600   ~~0.050~~

0.050   ....

> Place zeros on these numbers so they all have the same number of decimal places.

(1 mark)

4 Write these numbers in order, smallest first.

0.73   0.7   0.725   0.778   0.78

.... **(1 mark)**

Guided

5 Using the information that $5.7 \times 43 = 245.1$ write down the value of

(a) $57 \times 43 =$

$245.1 \times$ .... = ...

> 5.7 has been multiplied by 10 and 43 is unchanged. So 245.1 needs to be multiplied by 10.

(1 mark)

(b) $5.7 \times 4.3 =$

$245.1 \div$ .... = ...

> 5.7 is unchanged and 43 has been divided by 10. So 245.1 needs to be divided by 10.

(1 mark)

(c) $245.1 \div 57 =$

$43 \div$ .... = ...

> 245.1 is unchanged and 5.7 has been multiplied by 10. So 43 needs to be divided by 10.

(1 mark)

6 Sammy writes down the following in his exercise book.

$435.2 \div 13.6 = 320$

He uses the information that
$32 \times 136 = 4352$
Is he correct? Give a reason for your answer.

> Write 'yes' or 'no' and give your reason. You can use working to explain your answer if it is easier than writing it as a sentence.

.... **(1 mark)**

Operations on decimals

1 A coach ticket to the zoo costs £7.85. A teacher buys 36 of these tickets for his class.
What is the total cost of the 36 tickets?

> You can estimate the cost by first rounding both values to 1 significant figure and then multiplying.

Circle your answer.

£2826   £28 260   £282.60   £286.20   **(1 mark)**

2 Work out

> **Guided**

(a) 4.23 + 10.4

```
   4.23
+ 10.40
 ......3
```

> Make sure all the decimal points are lined up and then write zeros in the spaces so that all the numbers have the same number of decimal places.

(2 marks)

(b) 84.7 − 9.34

```
  84.70
−  9.34
 ......
```

(2 marks)

(c) 7.32 × 16

> First work out 732 × 16. In total there are 2 decimal places in the calculation, so put 2 decimal places in your answer.

.... **(2 marks)**

(d) 0.47 × 0.07

.... **(2 marks)**

(e) 83.4 ÷ 6

```
     ......
6)83.4
```

(2 marks)

(f) 81.9 ÷ 1.3

.... **(2 marks)**

3 Charles repairs computers.
He charged a customer £123.20 to repair a computer.
It took him eight hours to repair the computer.
How much did he charge for one hour?

£............................ **(2 marks)**

4 Kitty buys hot chocolate sachets.
There are 14 hot chocolate sachets in a small box.
A small box costs £3.49
Kitty uses three hot chocolate sachets each day.
Work out how much Kitty spends on hot chocolate sachets in a four-week period.

> For a longer question like this, it's a good idea to plan your strategy. Calculate:
> 1. number of days in a four-week period
> 2. number of sachets used in a four-week period
> 3. number of small boxes used in a four-week period
> 4. total cost of those boxes.

£............................ **(4 marks)**

Squares, cubes and roots

1 Work out

(a) 4^2

(b) 2^3

(c) $\sqrt{81}$

.... **(1 mark)** .... **(1 mark)** .... **(1 mark)**

(d) $\sqrt{64}$

(e) $\sqrt[3]{64}$

(f) $\sqrt[3]{-125}$

.... **(1 mark)** .... **(1 mark)** .... **(1 mark)**

2 Write down

(a) the square of 9

(b) the cube of 5

.... **(1 mark)** .... **(1 mark)**

(c) the square root of 144

(d) the cube root of 216

.... **(1 mark)** .... **(1 mark)**

3 Work out the value of $5^2 + 3^3$

> Square the 5 and cube the 3 before you add.

 Guided

$(5 \times 5) + (3 \times 3 \times 3) = $ .... $+$ .... $= $ .... **(1 mark)**

4 Work out the value of $5^3 - 6^2$

Circle your answer.

                 113   89   161   39   **(1 mark)**

5 2, 8, 11, 15, 21, 26, 36, 49

Write down a number from the list that

(a) is a square number

(b) is a cube number

(c) has a square root of 7

.... **(1 mark)** .... **(1 mark)** .... **(1 mark)**

6 Tom carried out an investigation and concluded that
     '6 is a cube number since $2^3 = 6$'

> You can explain your answer by writing a sentence with your reason, or by showing some neat working.

Is he correct? Give a reason for your answer.

 Guided

No, because $2 \times 2 \times 2 = $ .... **(1 mark)**

7

> If you add three square numbers then you always get an even number.

> One counter example (an example that does not work) is enough for a reason if the statement is incorrect.

Is this statement correct? Give a reason for your answer.

.... **(1 mark)**

Indices

1 Write as a single power of 4

Guided

   (a) $4 \times 4 = 4^{\cdots}$ **(1 mark)**   (b) $4 \times 4 \times 4 \times 4 \times 4 =$ .... **(1 mark)**

2 Simplify and leave your answers in index form.

Guided

   (a) $5^3 \times 5^6$ &boxed{Add the powers.}   (b) $5^9 \div 5^6$ &boxed{Subtract the powers.}

   $5^3 \times 5^6 = 5^{3+6} = 5^{\cdots}$ **(1 mark)** $5^9 \div 5^6 = 5^{9-6} = 5^{\cdots}$ **(1 mark)**

   (c) $\dfrac{5^{12}}{5 \times 5^7}$ &boxed{First work out the power of 5 in the denominator. Remember $5 = 5^1$}   (d) $(5^3)^4$ &boxed{Multiply the powers.}

   .... **(2 marks)**   .... **(1 mark)**

3 Simplify $4^5 \times 4^5$

   Circle your answer.

             4^{10} 16^{10} 4^{25} 16^{25} **(1 mark)**

4 Write as a single power of 9

Guided

   (a) $\dfrac{1}{9} = 9^{-}$ **(1 mark)**   (b) $\dfrac{1}{9 \times 9 \times 9 \times 9} =$ .... **(1 mark)**

5 Simplify and leave your answers in index form.

   (a) $\dfrac{8^2 \times 8^6}{8^5}$ (b) $\dfrac{8^{12}}{8^6 \times 8^4}$ (c) $\dfrac{8^7 \times 8^6}{8 \times 8^4}$ (d) $(8^4)^5$

   .... **(2 marks)**   .... **(2 marks)**   .... **(2 marks)**   .... **(1 mark)**

6 Work out &boxed{Anything to the power zero equals **one**.}

Guided

   (a) $7^0 =$ ....   **(1 mark)**   (b) $7^{-2} = \dfrac{1}{7^2} = \dfrac{1}{.....}$ **(1 mark)**

   (c) $4^{-3} =$ ....   **(1 mark)**   (d) $\left(\dfrac{3}{4}\right)^3 = \dfrac{3^3}{4^3} = \dfrac{.....}{.....}$ **(1 mark)**

7 $7^4 \times 7^x = \dfrac{7^9 \times 7^6}{7^3}$

   Work out the value of x

                                $x =$ .... **(2 marks)**

Estimation

1 Work out an estimate for the value of

> Round both values to 1 significant figure.

Guided

(a) $188 \times 69 \approx 200 \times 70 = $ .... **(1 mark)**

(b) $28.9 \div 4.85 \approx$ .... $\div$ .... = ... **(1 mark)**

2 Work out an estimate for the value of $\dfrac{4826}{4.1 \times 9.72}$

Guided

$\approx \dfrac{5000}{4 \times \text{.................}} = \dfrac{\text{.................}}{\text{.................}} = $ ....

> 1. Round all values to 1 significant figure.
> 2. Multiply the numbers in the denominator.
> 3. Cancel if possible, then divide.

(2 marks)

3 Work out an estimate for the value of $\dfrac{716 \times 5.13}{0.191}$

Guided

$\dfrac{700 \times 5}{0.2} = \dfrac{3500}{0.2} = \dfrac{\text{.............}}{2} = $ ....

> If you need to divide by a decimal you can multiply top and bottom by 10 or 100 to simplify the calculation.

(2 marks)

4 Work out an estimate for the value of $\dfrac{29 \times 4.90}{0.204}$

.... **(2 marks)**

5 Harry estimates the value of $\dfrac{38 \times 3.27}{0.53}$ to be 60

Harry's estimate is incorrect.

(a) Work out a correct estimate.

....

.... **(2 marks)**

(b) What mistake did Harry make?

.... **(1 mark)**

6 The radius of a sphere is 6.2 cm.

(a) Work out an estimate for the surface area of the sphere.

PROBLEM SOLVED!

> Surface area of a sphere $= 4\pi r^2$

> You will need to use problem-solving skills throughout your exam – **be prepared!**

> You can round π and r to 1 significant figure.

.... cm² **(2 marks)**

(b) Is your answer to part (a) an overestimate or an underestimate? Give a reason for your answer.

.... **(1 mark)**

Factors, multiples and primes

1 (a) Write down all the factors of 36

1 × 36, 2 × ...., ... × ...., ... × ...., ... × ....   **(2 marks)**

(b) Write down the first 10 multiples of 7

7   14   ....   **(1 mark)**

2 Here is a list of numbers:

2   8   6   12   21   25   33   49

From the list write down

(a) a factor of 30 ....   **(1 mark)**

(b) a multiple of 7 ....   **(1 mark)**

(c) 2 factors of 24 that have a product of 48 ....   **(2 marks)**

3 Write down three factors of 28 which have a sum between 20 and 25

> Start by listing the factors of 28.

.... **(2 marks)**

4 Which of these numbers has **exactly three** factors?
Circle your answer.

6   12   25   33   **(1 mark)**

5 The table shows some numbers.

Three of the numbers are prime numbers.
Put a tick (✓) underneath each of these three numbers.   **(1 mark)**

41	42	43	44	45	46	47	48	49

6 Write these numbers as products of their prime factors.
Give your answers in index form.

(a) 54

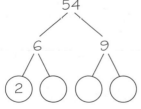

54 = 2 × .... × .... × .... = 2 × ....^....   **(3 marks)**

> The prime factors are always circled.

(b) 96   (c) 126   (d) 252

.... **(3 marks)**   .... **(3 marks)**   .... **(3 marks)**

HCF and LCM

1 (a) Work out the highest common factor (HCF) of 72 and 84

Guided

> 1. List the factors of 72.
> 2. List the factors of 84.
> 3. Circle all the common factors.
> 4. Choose the highest common factor.

1 × 72, 2 × ...., ... × ...., ... × ...., ... × ...., ... × ....

1 × 84, 2 × ...., ... × ...., ... × ...., ... × ...., ... × ....

.... **(3 marks)**

(b) Work out the lowest common multiple (LCM) of 12 and 15

.... **(2 marks)**

2 (a) Write these numbers as products of powers of their prime factors.

Guided

> Circle any prime numbers.

90

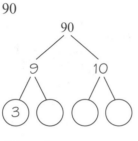

210

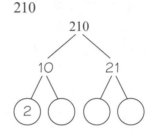

90 = 3 × .... × .... × .... **(2 marks)**   210 = 2 × .... × .... × .... **(2 marks)**

(b) Work out the highest common factor (HCF) of 90 and 210

90 = 3 × .... × .... × ....

210 = 2 × .... × .... × ....

HCF = ... × .... × .... = ...

> Circle all the prime numbers which are common to both products of prime factors. Multiply the circled numbers together to find the HCF.

(1 mark)

(c) Work out the lowest common multiple (LCM) of 90 and 210

LCM = ... × .... × .... = ...

> To find the LCM, multiply the HCF by the numbers in both products that were not circled in part (b).

(1 mark)

3 (a) Work out the highest common factor (HCF) of 36 and 48

.... **(2 marks)**

(b) Work out the lowest common multiple (LCM) of 36 and 48

.... **(2 marks)**

Fractions

 1 Shade $\frac{3}{7}$ of this shape.

(1 mark)

 2 Write these fractions in their simplest form.

 Guided (a) $\frac{30}{60} = \frac{\dots}{\dots}$ (1 mark)

(b) $\frac{12}{18} = \frac{\dots}{\dots}$ (1 mark)

(c) $\frac{35}{120}\begin{smallmatrix}\div 5 \\ \div 5\end{smallmatrix} = \frac{\quad}{\quad}$ What number will go into 35 and 120?

(d) $\frac{24}{84} = \frac{\dots}{\dots}$ (1 mark)

(1 mark)

 3 Write down the fraction of these shapes that are shaded.
Write your fraction in its simplest form.

(a) .... (2 marks)

(b) .... (2 marks)

 4 Work out

Guided (a) $\frac{3}{4}$ of £60

(b) $\frac{4}{5}$ of £80

 £60 ÷ 4 = ...

.... × 3 = £.................. (2 marks)

£.................. (2 marks)

(c) $\frac{7}{8}$ of £160

(d) $\frac{5}{7}$ of £210

£.................. (2 marks)

£.................. (2 marks)

 5 Which of these fractions is closest to $\frac{1}{3}$?

$\frac{7}{20}$ $\frac{4}{15}$ $\frac{3}{10}$ $\frac{5}{12}$

Show working to justify your answer.

....

....

.... (3 marks)

 6 Tom bought 20 boxes of flowers. Each box cost him £6
Each box contains 15 flowers.

He sells $\frac{3}{5}$ of the total number of flowers for 70p each.

He then sells the remaining flowers at 50p each.
Work out the total profit Tom makes.

£......................... (5 marks)

Operations on fractions

1 Work out

> Write both fractions as equivalent fractions with the same denominator.

(a) $\dfrac{1}{3} + \dfrac{2}{5}$

(b) $\dfrac{4}{5} - \dfrac{1}{4}$

$= \dfrac{5}{15} + \dfrac{\dots}{15} = \dfrac{\dots}{15}$ **(2 marks)**

$= \dfrac{\dots}{20} - \dfrac{\dots}{20} = \dfrac{\dots}{20}$ **(2 marks)**

(c) $\dfrac{6}{7} + \dfrac{3}{8}$

(d) $\dfrac{5}{9} - \dfrac{4}{7}$

.... **(2 marks)**   .... **(2 marks)**

2 Work out

(a) $\dfrac{1}{2} \times \dfrac{1}{3}$

(b) $\dfrac{5}{11} \times \dfrac{3}{4}$

.... **(1 mark)**   .... **(1 mark)**

(c) $\dfrac{4}{5} \div \dfrac{3}{10}$

> Turn the second fraction upside down and change ÷ into ×

(d) $\dfrac{2}{3} \div \dfrac{4}{9}$

.... **(2 marks)**   .... **(2 marks)**

3 A man wins some money and decides to give it to his three children.

Andrew receives $\dfrac{2}{5}$ of the money, Ben receives $\dfrac{1}{3}$ of the money and Carla receives the rest.
Work out the fraction that Carla receives.

$\dfrac{2}{5} + \dfrac{1}{3} = \dfrac{\dots}{15} + \dfrac{\dots}{15} = \dfrac{\dots}{15}$

$1 - \dfrac{\dots}{15} = \dfrac{\dots}{15}$

> Write 1 as a fraction with the same numerator and denominator. $1 = \dfrac{15}{15}$

(3 marks)

4 A garage has a supply of 210 litres of oil.

Amy uses $\dfrac{4}{7}$ of the supply and Brad uses $\dfrac{1}{5}$ of the supply.

(a) What fraction of the supply is left?

.... **(3 marks)**

(b) How much oil is left?

.... litres **(2 marks)**

Mixed numbers

1 Work out

Guided

(a) $3\frac{4}{5} + 2\frac{3}{4}$

> You need to write mixed numbers as improper fractions before you do any calculations.

$$= \frac{19}{5} + \frac{\ldots}{4} = \frac{\ldots}{20} + \frac{\ldots}{20} = \frac{\ldots}{20} = \ldots\ldots\ldots$$

> Write your final answer as a mixed number in its simplest form.

(3 marks)

(b) $4\frac{2}{5} - 2\frac{3}{10}$

$$- \frac{\ldots}{5} - \frac{\ldots}{10} = \frac{\ldots}{10} - \frac{\ldots}{10} = \frac{\ldots}{10} = \ldots\ldots\ldots$$

(3 marks)

2 Work out

Guided

(a) $1\frac{2}{3} \times 2\frac{3}{10}$

$$= \frac{\ldots}{3} \times \frac{\ldots}{10} = \frac{\ldots}{\ldots} = \ldots\ldots\ldots$$

(3 marks)

(b) $4\frac{2}{3} \div 1\frac{2}{5}$

> Don't forget to replace ÷ with × and then flip the fraction over.

$$= \frac{\ldots}{3} \div \frac{\ldots}{5} = \frac{\ldots}{3} \times \frac{\ldots}{\ldots} = \frac{\ldots}{\ldots} = \ldots\ldots\ldots$$

(3 marks)

3 Work out

(a) $3\frac{1}{2} \times 2\frac{4}{7}$ (b) $5\frac{1}{3} \div 1\frac{4}{9}$

.... **(3 marks)**   .... **(3 marks)**

4 It takes $4\frac{2}{3}$ hours to paint a room, and $1\frac{1}{4}$ hours for all the paint to dry.

How long does it take altogether?

.... hours **(3 marks)**

5 Andy needs six sections of piping, each $1\frac{4}{5}$ metres in length.

He has 10 metres of piping. Does he have enough?

You **must** show your working.

....

....

.... **(3 marks)**

15

Calculator and number skills

1 Work out

(a) $11 + 8 \div 2$

(b) $2 + 9 \times 10 + 3$

$11 +$ .... = ... **(1 mark)**

.... **(1 mark)**

(c) $8 + (3 \times 20) \div 6$

(d) $(14 - 5)^2$

.... **(1 mark)**

.... **(1 mark)**

2 Work out

You must use BIDMAS.

(a) $\dfrac{27 + 3 \times 3}{3 \times 2}$ .... **(1 mark)**

(b) $\dfrac{13 - 12 \div 4}{4 + 3 \times 2}$ .... **(1 mark)**

3 Work out $\dfrac{12 + 3 \times 6}{4 + 4 \div 2}$

Circle your answer.

22.5   15   7.5   5   **(1 mark)**

4 Work out the value of $\dfrac{4.5 + 3.75}{3.2^2 - 5.53}$

Write down all the figures on your calculator display.

$\dfrac{8.25}{\text{...................}}$ = ... **(2 marks)**

5 (a) Work out the value of $\sqrt{30.25} + 1.75^2$

Enter the numbers into the calculator. You might need to press the (SD) key to get your answer as a decimal number.

.... **(2 marks)**

(b) Write your answer to part (a) correct to 1 significant figure.

.... **(1 mark)**

6 (a) Work out the value of $\dfrac{\sqrt{18.3 + 3.6^2}}{2.8 \times 1.6}$

Write down all the figures on your calculator display.

.... **(2 marks)**

(b) Write your answer to part (a) correct to 3 significant figures.

.... **(1 mark)**

7 (a) Work out the value of $\dfrac{32.5 \times \sqrt[3]{16.3}}{9.5 \times 3.1}$

Write down all the figures on your calculator display.

.... **(2 marks)**

(b) Write your answer to part (a) correct to 2 significant figures.

.... **(1 mark)**

Standard form 1

1 (a) Write 45 000 in standard form.

$45\,000 = 4.5 \times 10^{\text{......}}$

> Count decimal places from the right. How many jumps do you need to make to get 4.5?

(1 mark)

(b) Write 3.4×10^{-5} as an ordinary number.

$3.4 \times 10^{-5} = 0.0\ldots\ldots$

> The power of 10 is negative so the number is less than 1.

(1 mark)

(c) Write 28×10^6 in standard form.

.... **(1 mark)**

2 Write in standard form

(a) 567 000   (b) 0.000 056 7   (c) 567×10^8

.... **(1 mark)**   .... **(1 mark)**   .... **(1 mark)**

3 (a) Write 6 740 000 in standard form.

.... **(1 mark)**

$n = 6\,740\,000$ and $m = 5.42 \times 10^5$
Work out, giving your answers in standard form correct to 2 significant figures.

> Use the $\boxed{\times 10^{x}}$ key to enter standard form numbers on your calculator.

(b) $n + m$

$6\,740\,000 + \ldots\ldots\ldots\ldots\ldots$

$= \ldots\ldots\ldots\ldots\ldots\ldots$ **(2 marks)**

(c) $n - m$

$6\,740\,000 - \ldots\ldots\ldots\ldots\ldots$

$= \ldots\ldots\ldots\ldots\ldots\ldots$ **(2 marks)**

4 In 2014 the population of the United Kingdom was 6.5×10^7
In 2014 the population of Russia was 1.4×10^8

Work out the difference between the population of the United Kingdom and the population of Russia. Give your answer in standard form.

.... **(2 marks)**

5 $p = 5 \times 10^4$ $q = 2 \times 10^7$

Work out $q \div p$
Circle your answer.

4×10^{-2} 4×10^2 2.5×10^{-3} 2.5×10^2 **(1 mark)**

6 In 2011, NASA launched the spacecraft Curiosity to land on the planet Mars. The distance from Earth to Mars is 225 million km. The time it took to reach Mars was 4320 hours.
Work out the average speed, in km/h, of the spacecraft Curiosity.
Give your answer in standard form correct to 2 significant figures.

$\text{speed} = \dfrac{\text{distance}}{\text{time}} = \ldots\ldots\ldots\ldots\ldots\ldots\ldots\ldots\ldots\ldots$ **(2 marks)**

Standard form 2

7 Work out, giving your answers in standard form

> Multiply the number parts then add the powers.

(a) $(3 \times 10^6) \times (6 \times 10^{-3})$

= $(3 \times \ldots\ldots) \times (10^6 \times 10^{\ldots}) = \ldots\ldots \times 10^{\ldots} = \ldots\ldots \times 10^{\ldots}$ **(2 marks)**

(b) $(8 \times 10^6) \div (4 \times 10^{-14})$

= $(8 \div \ldots\ldots) \times (10^6 \div 10^{\ldots}) = \ldots\ldots \times 10^{\ldots}$ **(2 marks)**

8 Work out, giving your answers in standard form

(a) $5.1 \times 10^3 + 6.5 \times 10^4$

```
   5100
+ 65000
```

(b) $7.6 \times 10^5 - 8 \times 10^3$

```
  760000
−   8000
```

.... **(2 marks)**   .... **(2 marks)**

9 A and B are standard form numbers.
$A = 5.6 \times 10^9$ $B = 8 \times 10^{-2}$
Work out, giving your answers in standard form

(a) $2A$ (b) $A \times B$ (c) $A \div B$

.... **(2 marks)**   .... **(2 marks)**   .... **(2 marks)**

10 It takes light 8 minutes to travel from the Sun to the Earth.
The speed of light is 3×10^8 m/s.
Work out the distance, in km, from the Sun to the Earth.
Give your answer in standard form.

> speed = $\dfrac{\text{distance}}{\text{time}}$

.... **(3 marks)**

11 The distance from the Sun to the planet Neptune is approximately 4.5×10^9 km.
The speed of light is 3×10^8 m/s.
Work out how long, in seconds, it takes light
to travel from the Sun to the planet Neptune.

> Convert the distance into metres,
> then use time = $\dfrac{\text{distance}}{\text{speed}}$

.... **(3 marks)**

Counting strategies

Guided

1   Ajay writes down one letter from the word ART then he writes down one number from 1, 2 and 3.

| A | R | T | | 1 | 2 | 3 |

Do not write down repeats such as (1, A) which is the same as (A, 1).

List all the possible combinations Ajay could write down.

(A,1) (A, ...) (A, ...) (R, ...) (R, ...) (R, ...)
(T, ...) (T, ...) (T, ...)

(2 marks)

2   Brett goes to a restaurant.
He can choose from three types of curry
and three types of naan.
Brett is going to choose one curry and one naan.
List all the possible combinations Brett can choose.

Curry	**Naan**
Chicken	Plain
Lamb	Garlic
Vegetable	Butter

.... **(2 marks)**

3   Emily has four tiles.
One tile is marked W, one tile is marked X,
one tile is marked Y and one tile is marked Z.
Emily chooses two of these tiles.
Write down all the possible combinations she can get.

| W | | X | | Y | | Z |

.... **(2 marks)**

4   Kate has three cards. Each card has a different digit on it.
Kate wants to make a three-digit number.
Each number is made with all three cards.
How many different numbers can Kate make?

| 3 | | 6 | | 9 |

.... **(2 marks)**

Guided

5   There are four players in a competition,
Asha, Bev, Chloe and Dan.
Each player must play each other once.
How many games will be played in total?

Label Asha, Bev, Chloe and Dan as A, B, C and D respectively.

Remember (A, B) is the same as (B, A).

(A, ...) (A, ...) (A, ...) (B, ...) ....

.... **(2 marks)**

6   Grace has five coins.   5p   10p   20p   50p   £1
She selects **two** coins at random.
How many different totals of money can she get?

....

.... **(2 marks)**

Problem-solving practice 1

1 Four different prime numbers are added together to make 35.
Write down two ways in which this can be done.

.... **(2 marks)**

2 Crisps cost 35p per packet. A bottle of lemonade costs £1.25. Nigel buys five packets of crisps and one bottle of lemonade. He pays with a £10 note.
Work out how much change he should get.

£... **(3 marks)**

3 Here are some properties of numbers: Odd   Prime   Square   2-digit

(a) Which **two** properties does the number 7 have?

.... **(1 mark)**

(b) Can one number have **all** of the properties?
Tick a box.

☐ Yes   ☐ No   ☐ Cannot tell

Give a reason for your answer.

.... **(1 mark)**

(c) Write down a number with **three** of the properties.
State which properties it has.

....

.... **(2 marks)**

4 Which of these fractions is closest to $\frac{2}{5}$?

$\frac{3}{8}$ $\frac{5}{12}$ $\frac{7}{20}$ $\frac{17}{40}$ **(1 mark)**

Show working to justify your answer.

....

....

.... **(3 marks)**

5 A machine makes 48 bolts every hour. The machine makes bolts for $7\frac{1}{2}$ hours each day, on five days of the week. The bolts are packed into boxes. Each box holds 30 bolts. How many boxes are needed for all the bolts made each week?

....boxes **(4 marks)**

6 Tammy buys three compost bags. Each compost bag has a mass of 24 kg.
She can fill a small pot by using $\frac{4}{9}$ of a compost bag.
How many pots can she fill?

.... **(3 marks)**

Problem-solving practice 2

7   Here are four boxes containing counters.

| Box A | | Box B | | Box C | | Box D |

There are 114 counters in total.

The total number of counters in boxes A and B is 54

Box C has twice as many counters as Box D.

The difference between the number of counters in Box A and Box D is 11

Box C has the greatest number of counters.

Work out the number of counters in each box.

....

....

....

.... **(4 marks)**

8   A plumber has copper pipes of lengths 54 cm and 72 cm.

He wants to cut them and make smaller pipes to use them in boilers.

He wants all the smaller pipes to be the same length with no copper left over.

What is the greatest length of the smaller pipes?

....cm   **(3 marks)**

9   Buses leave Wolverhampton to Penn every 12 minutes and to Wombourne every 15 minutes.

A bus to Penn and a bus to Wombourne leave Wolverhampton at 8 am. At what time will a bus to Penn and a bus to Wombourne next leave Wolverhampton at the same time?

.... **(3 marks)**

10   An atomic particle has a lifetime of 4.86×10^{-5} seconds.

It travels at a speed of 6.2×10^4 m/s. Show that the distance travelled by the atomic particle is approximately 3 m.

(3 marks)

Collecting like terms

1 Simplify

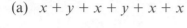

(a) $x + x + x + x + x$

(b) $3w + 5w - 2w$

= ....x **(1 mark)**

= ....w **(1 mark)**

2 Simplify

(a) $x + y + x + y + x + x$

= ....$x +$ ....y **(1 mark)**

(b) $8ab - 3ab$

= ....ab **(1 mark)**

(c) $5t + 6v - 4t + 5v$

= $5t - 4t + 6v + 5v =$ ....$t +$ ....v **(2 marks)**

(d) $6c + 2d - 3c - 4d$

= $6c - 3c + 2d - 4d =$ ....$c -$ ....d **(2 marks)**

3 Simplify

(a) $5x - 3x$

(b) $4t^2 - t^2$

.... **(1 mark)**

.... **(1 mark)**

(c) $4a + 5b - a + 3b + 7$

(d) $6x - 3y - 5x - 4y$

.... **(2 marks)**

.... **(2 marks)**

(e) $6p + 8q + 2q - 9q$

(f) $5 + 3a + 7b - a - b$

.... **(2 marks)**

.... **(2 marks)**

4 Simplify $9m - 4n - 2m + n$

Circle your answer.

$5m - n$ $7m - 5n$ $4mn$ $7m - 3n$ **(1 mark)**

5 formula equation expression term

Choose a word from the list above to describe each of the following.

(a) $3x + y$ (b) $v = u + at$ (c) $3x + 4 = 6$

.... **(1 mark)** .... **(1 mark)** .... **(1 mark)**

Simplifying expressions

1 Simplify

(a) $y \times y$

$= y^{......}$ **(1 mark)**

(b) $3m \times t$

$= 3 \times m \times t = $ .... **(1 mark)**

2 Simplify

(a) $w \times w \times w \times w$

.... **(1 mark)**

(b) $4 \times 7 \times d$

.... **(1 mark)**

(c) $5 \times 6k$

.... **(1 mark)**

(d) $5j \times 8k$

.... **(1 mark)**

3 Simplify

(a) $5x \times 3x$

$= 5 \times 3 \times x \times x = $ .... **(1 mark)**

(b) $2e \times 3f$

$= 2 \times 3 \times e \times f = $ .... **(1 mark)**

(c) $8a \div 2$ [Work out $8 \div 2$.]

$= \dfrac{8a}{2} = $ .... **(1 mark)**

(d) $24ab \div 3a$ [Divide the number parts and then the letter parts.]

$= \dfrac{24ab}{3a} = $ .... **(1 mark)**

4 Simplify

(a) $7g \times 5h$

.... **(1 mark)**

(b) $2t \times 2t \times 2t$

.... **(1 mark)**

(c) $36xy \div 12y$

.... **(1 mark)**

(d) $28xyz \div 7xz$

.... **(1 mark)**

5 Simplify

(a) $2a \times 3b \times 4c$

.... **(1 mark)**

(b) $48mnp \div 6mn$

.... **(1 mark)**

6 **Two** of these expressions are equivalent to $2xy$

$4x \div 2y$ ☐ $2(x \times y)$ ☐ $2x \times 2y$ ☐ $4xy \div 2$ ☐ $x \times y$ ☐

Tick (✓) the boxes next to the correct expressions. **(2 marks)**

Algebraic indices

1 Simplify

(a) $a^3 \times a^6$ Add the powers.

.... **(1 mark)**

(b) $a^9 \div a^6$ Subtract the powers.

.... **(1 mark)**

(c) $\dfrac{a^{12}}{a \times a^7}$ First work out the power of a in the denominator.

.... **(1 mark)**

(d) $(a^3)^4$ Multiply the powers.

.... **(1 mark)**

2 Simplify

(a) $\dfrac{t^2 \times t^6}{t^5}$

.... **(1 mark)**

(b) $\dfrac{t^{12}}{t^5 \times t^4}$

.... **(1 mark)**

(c) $\dfrac{t^7 \times t^6}{t \times t^4}$

.... **(2 marks)**

(d) $\dfrac{t^8 \times t^{-6}}{t \times t^{-5}}$

.... **(2 marks)**

3 Simplify

> Guided

(a) $(x^3)^4$

.... **(1 mark)**

(b) $(4x^2)^3$

$= 4x^2 \times 4x^2 \times 4x^2 =$ .... **(2 marks)**

(c) $(2x^3)^3$

.... **(2 marks)**

(d) $3x^2 \times 4x^5$

$= 3 \times$ .... $\times x^{....\, +\, ...} =$ .... **(2 marks)**

(e) $3x^2y \times 4x^5y^4$

.... **(2 marks)**

(f) $18x^3y^5 \div 6xy^2$

.... **(2 marks)**

4 Work out the value of x

(a) $p^4 \times p^x = p^{12}$

(b) $p^{12} \div p^x = p^7$

(c) $(p^3)^x = p^{15}$

$x =$ .... **(1 mark)** $x =$ .... **(1 mark)** $x =$ .... **(1 mark)**

5 Work out the value of x

$$q^5 \times q^{2x} = \dfrac{q^{10} \times q^6}{q^4}$$

Circle your answer.

2   1.5   3.5   7   **(1 mark)**

Substitution

1 Work out the value of

> Use BIDMAS to find the correct value.

(a) $5x + 3$ when $x = 4$

(b) $2x - 3$ when $x = -4$

$= 5 \times \text{.......} + 3 = \text{.......}$ **(2 marks)**

$= 2 \times \text{.......} - 3 = \text{.......}$ **(2 marks)**

2 Work out the value of

(a) $4a + 3b$ when $a = 4$ and $b = 6$

(b) $3a - 5b$ when $a = 3$ and $b = -2$

.... **(2 marks)**

.... **(2 marks)**

3 Which of these expressions has the smallest value when $x = 8$ and $y = -2$?

$$x - y \qquad\qquad xy \qquad\qquad \frac{x}{y}$$

You **must** show your working.   **(2 marks)**

4 Work out the value of

(a) $3a + ax$ when $a = 3$ and $x = -5$

$= 3 \times \text{.........} + \text{.........} \times \text{.........} = \text{...}$

> $+ \times - = -$

(2 marks)

(b) $4t^2$ when $t = -5$

$= 4 \times (\text{.........})^2 = 4 \times \text{.........} \times \text{.........} = \text{...}$

> $- \times - = +$

(2 marks)

(c) $3g^2 - 5g$ when $g = 3$

$= 3 \times (\text{.........})^2 - 5 \text{.........} = \text{...}$ **(2 marks)**

5 Work out the value of

(a) $4(2x - 4y)$ when $x = 3$ and $y = -5$

.... **(2 marks)**

(b) $3m + 5(p - n)$ when $m = 6$, $n = 2$ and $p = 3$

.... **(2 marks)**

(c) $9t - \frac{1}{2}at^2$ when $a = 2$ and $t = 4$

.... **(2 marks)**

6 Abbie and Lisa are trying to work out the energy of a ball when it is dropped.
They use the following formula: $E = \frac{1}{2}mv^2$ where $m = 2$ and $v = 3$
Abbie works out the value of E to be 9
and Lisa works out the value of E to be 18.
Who is correct? Give a reason for your answer.

> You will need to show some working to justify your answer.

.... **(2 marks)**

Formulae

1 The time in minutes needed to cook a joint of meat is given by the formula in the box. Work out the time needed to cook a 5 kg joint of meat.

> **Guided**

Substitute the value for the mass of the joint into the formula.

Time = Mass in kg × 15 + 30

Use BIDMAS to find the correct value.

Time = ... × 15 + 30 = ... hour .... minutes   **(2 marks)**

2 Andy the carpenter charges £25 for each hour he works at a job plus a £50 callout charge. The amount Andy charges, in pounds, can be worked out using the formula in the box. Andy works for six hours at a job. Work out how much Andy charges.

Charge = Number of hours worked × 25 + 50

£........................... **(2 marks)**

3 The height h cm of a growing tree is given by the formula $h = 3t + 12$
Work out the value of h when $t = 6$, where t is the time in days.

> **Guided**

Substitute the value of t into the formula.

$h = 3 \times$ .... + ... = ...   **(2 marks)**

4 A formula involving force, mass and acceleration is $F = ma$
Work out the value of F when $m = 12$ and $a = 3$

$F =$ .... **(2 marks)**

5 This formula is used to calculate impulse: $I = mu - mv$
Work out the value of I when $m = 6$, $u = 8$ and $v = 5$

$I =$ .... **(2 marks)**

6 A formula to work out the velocity of a ball is $v = u + at$
Work out the value of v when $u = -20$, $a = 9$ and $t = 8$

Substitute the values u, a and t into the formula.

(2 marks)

7 You can use this formula to convert degrees Celsius, C, into degrees Fahrenheit, F.

$F = 1.8C + 32$

(a) Use the formula to convert $-20\,°C$ into $°F$.

........................$°F$ **(2 marks)**

(b) One of these temperatures is the same in $°C$ and $°F$.
Circle the correct temperature.

$-50°$ $-40°$ $-30°$ $-10°$ **(1 mark)**

Writing formulae

1 A can of lemonade costs g pence. A bag of sweets cost h pence.
Harry buys four cans of lemonade and five bags of sweets.
Write a formula, in terms of g and h, for the total cost, C pence.

$C = 4 \times$ .... $+ 5 \times$ .... $= 4$ .... $+ 5$ .... **(2 marks)**

2 Aliya plays a game with black counters and white counters. The number of points
for a black counter is 10 and the number of points for a white counter is 20.
Aliya has m black counters and n white counters.
Her total number of points is S.
Write a formula for S in terms of m and n.

> Your formula should start $S =$

.... **(2 marks)**

3 Jayne the electrician charges £30 for each hour she works at a job and a callout
charge of £50. Jayne works n hours at a job. She charges P pounds.
Write a formula for P in terms of n.

$P = n \times$ .... $+$ .... $=$ .... $+$ .... **(2 marks)**

4 The cost, of £T, of hiring a taxi is a fixed £4 fee plus £3 for every kilometre travelled.
Which of these represents a formula for the cost of travelling x kilometres?

Circle your answer.

$\quad\quad T = 7x \quad\quad\quad T = 3 + 4x \quad\quad T = 4 + 3x \quad\quad T = 7 + 3x$ **(1 mark)**

5 The diagram shows the lengths of the sides of a quadrilateral. All the lengths are in cm.
Write a formula, in terms of x, for the perimeter, P cm, of the quadrilateral.

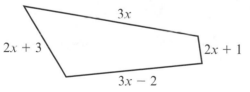

> The perimeter is the sum of
> the lengths of all the sides.

$P = 2x + 3 +$ .... $+$ .... $+$ .... $=$ .... **(2 marks)**

6 Alex is n years old. Brad is four years older than Alex.

(a) Write a formula, in terms of n, for Brad's age, B.

.... **(1 mark)**

The combined age of Alex, Brad and Carl is $5n + 4$

(b) Is Carl three times older than Alex? Give a reason for your answer.

.... **(2 marks)**

Expanding brackets

1 Expand

(a) $3(x + 2)$ **Guided**

$= 3 \times x + 3 \times 2$

$= \ldots\ldots + \ldots\ldots$ **(2 marks)**

(b) $4(x + 5)$

$\ldots\ldots\ldots\ldots\ldots\ldots\ldots\ldots$ **(2 marks)**

(c) $5(x - 3)$

$\ldots\ldots\ldots\ldots\ldots\ldots\ldots\ldots$ **(2 marks)**

(d) $6(2x + 3)$

$= 6 \times \ldots\ldots + 6 \times \ldots\ldots$

$= \ldots\ldots + \ldots\ldots$ **(2 marks)**

(e) $5(1 - 4x)$

$\ldots\ldots\ldots\ldots\ldots\ldots\ldots\ldots$ **(2 marks)**

(f) $7(3x - 8)$

$\ldots\ldots\ldots\ldots\ldots\ldots\ldots\ldots$ **(2 marks)**

2 Expand

> Multiply everything inside the brackets by **negative** 3. $-3 \times -3 = 9$ so you need to **add** 9 in part (a).

(a) $-3(x - 3)$ **Guided**

$= -3 \times x - 3 \times -3$

$= \ldots\ldots + \ldots\ldots$ **(2 marks)**

(b) $-4(x + 3)$

$\ldots\ldots\ldots\ldots\ldots\ldots\ldots\ldots$ **(2 marks)**

(c) $-6(x - 5)$

$\ldots\ldots\ldots\ldots\ldots\ldots\ldots\ldots$ **(2 marks)**

(d) $-2(2x + 3)$

$= -2 \times \ldots\ldots + -2 \times \ldots\ldots$

$= \ldots\ldots + \ldots\ldots$ **(2 marks)**

(e) $-2(4x - 1)$

$\ldots\ldots\ldots\ldots\ldots\ldots\ldots\ldots$ **(2 marks)**

(f) $-(2x - 4)$

$\ldots\ldots\ldots\ldots\ldots\ldots\ldots\ldots$ **(2 marks)**

3 Expand

(a) $x(x + 1)$ **Guided**

$= x \times x + x \times 1$

$= \ldots\ldots + \ldots\ldots$ **(2 marks)**

(b) $x(x + 5)$

$\ldots\ldots\ldots\ldots\ldots\ldots\ldots\ldots$ **(2 marks)**

(c) $2x(x - 9)$

$\ldots\ldots\ldots\ldots\ldots\ldots\ldots\ldots$ **(2 marks)**

(d) $3x(2x - 3)$

$= 3x \times \ldots\ldots + 3x \times \ldots\ldots$

$= \ldots\ldots + \ldots\ldots$ **(2 marks)**

(e) $-x(2x - 3)$

$\ldots\ldots\ldots\ldots\ldots\ldots\ldots\ldots$ **(2 marks)**

(f) $-3x(4x - 5)$

$\ldots\ldots\ldots\ldots\ldots\ldots\ldots\ldots$ **(2 marks)**

4 Expand and simplify

(a) $4x + 3(x + 2)$ **Guided**

$= 4x + 3 \times x + 3 \times 2$

$= \ldots\ldots x + \ldots\ldots x + \ldots\ldots$

$= \ldots\ldots x + \ldots\ldots$ **(3 marks)**

(b) $2(x + 1) + 3(x + 4)$

$= 2 \times x + 2 \times 1 + 3 \times x + 3 \times 4$

$= \ldots\ldots x + \ldots\ldots x + \ldots\ldots + \ldots\ldots$

$= \ldots\ldots x + \ldots\ldots$ **(3 marks)**

(c) $5(x - 3) + 4(2x + 1)$

$\ldots\ldots\ldots\ldots\ldots\ldots\ldots\ldots\ldots\ldots\ldots\ldots\ldots\ldots\ldots\ldots$ **(3 marks)**

(d) $4x(x - 3) + 2x(x - 4)$

$\ldots\ldots\ldots\ldots\ldots\ldots\ldots\ldots\ldots\ldots\ldots\ldots\ldots\ldots\ldots\ldots$ **(3 marks)**

5 Tom expands $5x(2 + 3x) - 4x(3x - 2)$

He gets the answer $2x + 3x^2$

Is Tom correct?

Show working to support your answer.

(3 marks)

Factorising

1 Factorise

| Expand your answer to check that it gives the original expression. |

(a) $3x + 6$

(b) $6a + 18$

(c) $2p - 6$

$= 3(.... + ...)$ **(1 mark)**

.... **(1 mark)**

.... **(1 mark)**

(d) $5y - 15$

(e) $3t + 24$

(f) $4g - 20$

.... **(1 mark)**

.... **(1 mark)**

.... **(1 mark)**

2 Factorise

(a) $x^2 + 6x$

(b) $x^2 - 4x$

(c) $x^2 - 9x$

$= x(.... + ...)$ **(1 mark)**

.... **(1 mark)**

.... **(1 mark)**

(d) $x^2 - 12x$

(e) $x^2 + 5x$

(f) $x^2 - x$

.... **(1 mark)**

.... **(1 mark)**

.... **(1 mark)**

3 Factorise fully

| 'Factorise fully' means that you need to take out the highest common factor. |

(a) $3p^2 + 6p$

(b) $8y^2 - 24y$

(c) $9t^2 - 36t$

$= 3p(.... + ...)$ **(2 marks)**

.... **(2 marks)**

.... **(2 marks)**

(d) $4d^2 + 12d$

(e) $6x^2 - 18x$

(f) $7n^2 - 35n$

.... **(2 marks)**

.... **(2 marks)**

.... **(2 marks)**

PROBLEM SOLVED!

4 Here are some factors.

| $2x^2 - 3xy$ 3 $3xy$ $3x$ $2x - 3y$ $2x^2 - 3y$ |

You will need to use problem-solving skills throughout your exam – **be prepared!**

Circle all the factors of $6x^2 - 9xy$ shown here.

(2 marks)

5 Match each expression A, B, C and D with its correct factorisation 1, 2, 3 or 4 **(3 marks)**

(A) $9x^2 - 15xy$

(B) $9xy - 15x^2$

(C) $9y^2 - 15xy$

(D) $9xy - 15y^2$

(1) $3y(3y - 5x)$

(2) $3y(3x - 5y)$

(3) $3x(3y - 5x)$

(4) $3x(3x - 5y)$

Linear equations 1

1 Solve

(a) $2x = 32$ $(\div 2)$

(b) $3x = -15$

(c) $-30 = 2q$

$x = $ .... **(1 mark)**

.... **(1 mark)**

.... **(1 mark)**

(d) $\dfrac{v}{-4} = 9$ $(\times -4)$

(e) $10 = \dfrac{x}{-12}$

(f) $72 = -8n$

$v = $ .... **(1 mark)**

.... **(1 mark)**

.... **(1 mark)**

2 Solve

(a) $x + 4 = 9$ (-4)

(b) $20 = p + 8$

(c) $7 - t = 5$

$x = $ .... **(1 mark)**

.... **(1 mark)**

.... **(1 mark)**

(d) $-2 = a - 5$

(e) $h - 4 = 15$

(f) $6 = -k - 40$

The = sign is symmetrical. You can swap the right- and left-hand sides of the equation.

$a - 5 = -2$ $(+ 5)$

$a = $ .... **(1 mark)**

.... **(1 mark)**

.... **(1 mark)**

3 Solve

(a) $2x - 8 = 4$ $(+8)$

(b) $7p + 30 = 9$

$2x = $ .... $(\div 2)$

$x = $ .... **(2 marks)**

.... **(2 marks)**

(c) $4t + 25 = 9$

(d) $2 = 3f + 14$

.... **(2 marks)**

.... **(2 marks)**

(e) $4 + \dfrac{h}{3} = 9$ (-4)

(f) $-2 = \dfrac{c}{3} + 6$

$\dfrac{h}{3} = $ .... $(\times 3)$

$h = $ .... **(2 marks)**

.... **(2 marks)**

4 Sam chooses three different numbers.
The second number is four times the first number.
The third number is six less than the first number.
The sum of the three numbers is 48.
Use algebra to work out the three numbers.

You will need to use problem-solving skills throughout your exam – **be prepared!**

(4 marks)

Linear equations 2

Guided

5 Solve

(a) $6x + 3 = 2x + 11$ $(-2x)$

$4x + 3 = 11$ (-3)

$4x = \ldots\ldots\ldots$ $(\div 4)$ │ Collect the x terms on one side. │

$x = \ldots\ldots\ldots$ **(3 marks)**

(b) $5x + 4 = 3x - 12$ (c) $7t - 12 = 3t - 9$

\ldots\ldots\ldots\ldots\ldots\ldots **(3 marks)**   \ldots\ldots\ldots\ldots\ldots\ldots **(3 marks)**

6 Solve   │ Start by multiplying out the brackets. │

(a) $3(2x - 1) = 27$ (b) $4(2x - 1) = 3x + 6$

\ldots\ldots\ldots\ldots\ldots\ldots **(3 marks)**   \ldots\ldots\ldots\ldots\ldots\ldots **(3 marks)**

(c) $2(6 - x) = 3(2x + 12)$ (d) $\dfrac{4x + 8}{5} = 4$

\ldots\ldots\ldots\ldots\ldots\ldots **(3 marks)**   \ldots\ldots\ldots\ldots\ldots\ldots **(3 marks)**

7 Liz is solving this equation: $3(4x - 5) = 2x + 9$

This is her solution: $12x - 5 = 2x + 9$
                             $12x - 2x = 9 - 5$
                             $10x = 4$
                             $x = 0.4$

Is Liz correct?
You **must** show your working.

(3 marks)

Number machines

1 Here is a number machine.

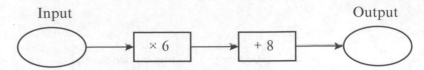

The output is twice the input.

(3 marks)

Work out the input.

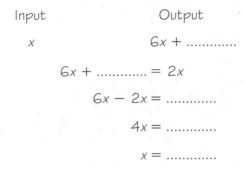

Input
x

Output
6x + ...

6x + ... = 2x

6x − 2x = ...

4x = ...

x = ...

2 Here is a number machine.

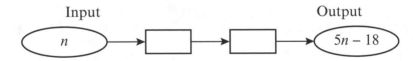

The input is n
The output is $5n - 18$

(a) Fill in the boxes of the number machines so that they represent
this two-stage operation.

(1 mark)

(b) Work out the value of n when the input and the output are equal.

(2 marks)

3 Here are two number machines A and B.

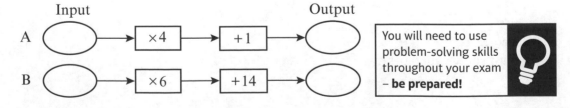

You will need to use
problem-solving skills
throughout your exam
– **be prepared!**

Both machines have the same input.

Work out the input that makes the output of **B** **ten greater than** the output of A.

(4 marks)

Inequalities

1 Write the inequalities shown on each number line.

Guided

> Closed (filled in) circles show numbers **are** included.

(a)
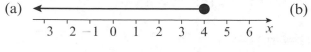

$x \leqslant$ .... **(1 mark)**

(b)

.... **(1 mark)**

(c)

.... $< x <$ .... **(1 mark)**

(d)

.... **(1 mark)**

2 Show each inequality on the number line.

> Use an open circle to show if a number is **not** included.

(a)

$x \geqslant 3$ **(1 mark)**

(b)

$x < 4$ **(1 mark)**

(c)

$-2 < x \leqslant 6$ **(1 mark)**

(d)

$-1 < x < 5$ **(1 mark)**

3 x is an integer. Write all the possible values of x

Guided

> Integer means whole number.

> $-3 <. x$ and $x \leqslant 4$

(a) $-3 < x \leqslant 4$

$x = -2,$ ...., ..., ..., ..., ..., ... **(1 mark)**

(b) $-2 \leqslant x < 3$

.... **(1 mark)**

(c) $-4 < x < 2$

.... **(1 mark)**

4 (a) p and q are **whole** numbers such that $p > 50$ and $q < 35$

Work out the **smallest** possible value of $p - q$ **(2 marks)**

(b) h and k are **whole** numbers such that $h < 45$ and $k \leqslant 28$

Work out the **largest** possible value of $h + k$ **(2 marks)**

Solving inequalities

1 Solve

(a) $2x \leqslant 20$ $(\div 2)$

$x \leqslant$ .... **(1 mark)**

(b) $3x > 15$

.... **(1 mark)**

(c) $4x \geqslant 16$

.... **(1 mark)**

(d) $3x \leqslant -16$

$x \leqslant$ .... **(1 mark)**

(e) $5x - 10 > 0$

.... **(1 mark)**

(f) $6x + 4 \geqslant 0$

.... **(1 mark)**

2 Solve

(a) $3x + 1 \geqslant 19$ (-1)

$3x \geqslant$ .... $(\div 3)$

$x \geqslant$ .... **(2 marks)**

(b) $5x - 8 < 27$

.... **(2 marks)**

(c) $5x - 12 > x$

.... **(2 marks)**

(d) $4x + 6 \leqslant 2x$

.... **(2 marks)**

3 x is an integer.
Write down all the possible values of x

'Integer' means 'whole number'.

(a) $-6 < 2x \leqslant 2$

.... **(3 marks)**

(b) $-5 < 3x \leqslant 13$

.... **(3 marks)**

4 Work out the integer value of x that satisfies both the inequalities.

$2x - 3 > 5$ and $3x + 4 < 22$

Solve both inequalities first.

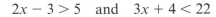

.... **(3 marks)**

5 Simon chooses three numbers.
The first number is a square number.
The second number is three times the first number.
The third number is seven more than the first number.
The sum of the three numbers is greater than 70 but less than 100.

You will need to use problem-solving skills throughout your exam – **be prepared!**

Work out the three numbers.

(5 marks)

Sequences 1

1   Here are some patterns made from sticks.

Pattern number 1   Pattern number 2   Pattern number 3   Pattern number 4

    (a)   Draw pattern number 4 in the space above.   **(1 mark)**

    (b)   How many sticks are in pattern 5?   **(1 mark)**

    (c)   Kate says that 1 of the patterns will be made from 30 sticks.
Is Kate correct? Give a reason for your answer.   **(1 mark)**

2   Write the next **two** terms for each of these number sequences.

 Guided

    (a)   2   6   10   14   ....   **(2 marks)**

    (b)   3   8   13   18   ....   **(2 marks)**

    (c)   1   3   9   27   ....   **(2 marks)**

    (d)   1   4   9   16   ....   **(2 marks)**

3   The rule for generating this sequence is 'add two
consecutive terms to get the next term'.

1 + ... = 4

1   4   7   ....

Work out the three missing numbers.   **(3 marks)**

4   The terms in this sequence decrease by the same amount each time.

Guided

    −4

38   34   30   26

    (a)   Write the next **two** terms in this sequence.

                ....   **(2 marks)**

    (b)   Ravina says that seven is a number in this sequence. Is she correct? Give a reason for your answer.

....   **(1 mark)**

5   Here is a sequence.
The third term of the sequence is 11.

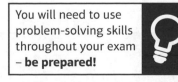
You will need to use problem-solving skills throughout your exam – **be prepared!**

....   11   ....

The rule for this sequence is
'add four to previous term then divide by two'.
Work out the first term of the sequence.

PROBLEM SOLVED!

                                          ....   **(3 marks)**

Sequences 2

6 Here are some linear sequences. Work out an expression for the *n*th term of each sequence.

(a) +4 +4 +4

  5   9   13   17

*n*th term = 4*n* .... **(2 marks)**

(b) 2   5   8   11

.... **(2 marks)**

(c) 2   9   16   23

.... **(2 marks)**

(d) 8   13   18   23

.... **(2 marks)**

7 Here are the first five terms of a linear sequence.

  4   7   10   13   16

Work out an expression, in terms of *n*, for the *n*th term of the sequence.

.... **(2 marks)**

8 Here are the first five terms of an arithmetic sequence.

3   7   11   15   19

(a) Work out an expression, in terms of *n*, for the *n*th term of the sequence.

.... **(2 marks)**

(b) Molly says that 199 is a term in the arithmetic sequence. Is Molly correct? Give a reason for your answer.

....

> Set the *n*th term of the sequence equal to 199 and solve the equation to find *n*.
> The term is part of the sequence only if *n* is an integer.

(2 marks)

9 The *n*th term of a sequence is $4n - 3$

The *n*th term of a different sequence is $60 - 3n$

Work out the **three** numbers that are in both sequences **and** between 30 and 60.

> You will need to use problem-solving skills throughout your exam – **be prepared!**

...................., ..., ... **(3 marks)**

Coordinates

1 Write the coordinates of

(a) point A

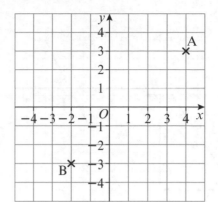

   (4, ...)   **(1 mark)**

(b) point B.

   (..........., ...)   **(1 mark)**

Plot the point

(c) (4 , −2) and label it C   **(1 mark)**

(d) (−1, 3) and label it D   **(1 mark)**

(e) What type of quadrilateral is ACBD?   **(1 mark)**

2 Work out the midpoints for the line segments given by these pairs of coordinates.

(a) (3, 6) and (7, 12)

Midpoint $= \left(\dfrac{3+7}{2}, \dfrac{\text{.............} + \text{.............}}{2}\right) = \left(\text{................}, \text{................}\right)$ **(2 marks)**

(b) (1, 8) and (9, 3)   (c) (4, 7) and (−8, 13)   (d) (2, −6) and (−10, 12)

....

   (2 marks)   **(2 marks)**   **(2 marks)**

3 Work out the midpoints of these line segments.

(a) 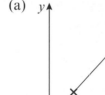   (b)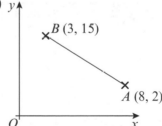

.... **(2 marks)**

4 Two straight lines are shown.
   A is the midpoint of *OB* and *B* is the midpoint of *PQ*
   Show that the coordinates of *P* are (2, 10)

   Start by working out
   the coordinates of *B*.

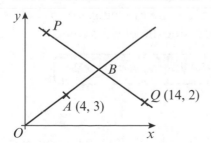

                                                                                      (2 marks)

Gradients of lines

1

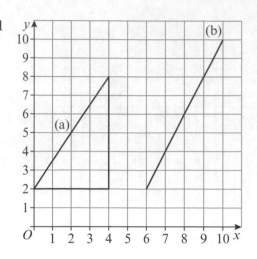

> **Guided**

Work out the gradients of the straight lines shown on the grid.

> Draw a triangle on the graph and use it to find the gradient.

(a) Gradient $= \dfrac{\text{distance up}}{\text{distance across}}$

$= \dfrac{\text{...........}}{\text{..........}}$

$= \text{..........}$ **(2 marks)**

(b)

.... **(2 marks)**

2

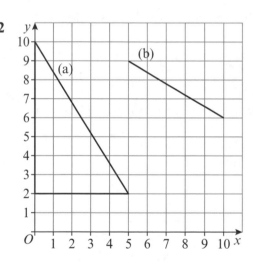

> **Guided**

Work out the gradients of the straight lines shown on the grid.

> The lines slope **down** so the gradient is **negative**.

(a) Gradient $= \dfrac{\text{distance up}}{\text{distance across}}$

$= \dfrac{\text{...........}}{\text{..........}}$

$= \text{..........}$ **(2 marks)**

(b)

.... **(2 marks)**

3

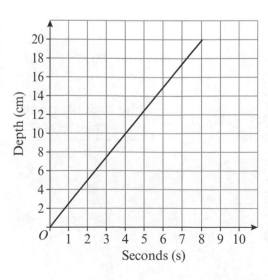

> The gradient of the line gives the rate of change.

Kim pours diesel into a container. The graph shows how the depth, in cm, of the diesel changes with time, in seconds.
Show that the rate of change of the depth of the diesel is 2.5 cm/s.

> Look carefully at the vertical scale.

.... **(2 marks)**

Straight-line graphs 1

Guided

1

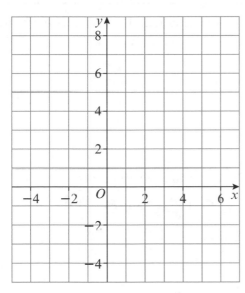

The x- and y-axes have different scales. Take care when plotting the points.

(a) Complete the table of values for $y = 2x - 1$

x	-2	-1	0	1	2	3
y		-3				5

> Substitute each value for x into the rule $y = 2x - 1$ to find the value of y.

$x = -2$: $y = (2 \times -2) - 1$

$= \ldots\ldots\ldots - 1 = \ldots\ldots\ldots$

$x = 1$: $y = (2 \times \ldots\ldots\ldots) - 1$

$= \ldots\ldots\ldots - 1 = \ldots\ldots\ldots$ **(2 marks)**

(b) On the grid draw the graph of $y = 2x - 1$

    (2 marks)

2

On the grid draw the graph of $x + y = 5$ for values of x from -3 to 6.

> First make a table of values. The question tells you to use 'values of x from -3 to 6'. Next work out the values of y.

    (3 marks)

Guided

3

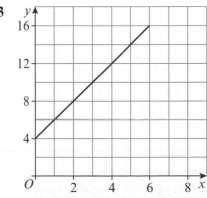

Use $y = mx + c$ to find the equation of a straight line.
c is the intercept on the y-axis.

Work out the equation of the straight line.

> Draw a triangle on the graph and use it to find the gradient m.

$\text{Gradient} = \dfrac{\text{distance up}}{\text{distance across}} = \dfrac{\ldots\ldots\ldots}{\ldots\ldots\ldots}$

$= \ldots\ldots\ldots\ldots\ldots$

$y = \ldots\ldots\ldots x + \ldots\ldots\ldots$ **(3 marks)**

Straight-line graphs 2

4 Circle the equation of a line that is parallel to $y = 6x - 4$

$y = 4x - 6$ $y = 6x + 3$ $y = 3x + 6$ $y = -\dfrac{1}{2}x - 4$ **(1 mark)**

5 Work out the equation of the straight line with

> **Guided**

(a) gradient 3, passing through the point (2, 5)

(b) gradient -2, passing through the point (3, 6)

$y = 3x + c$

$5 = $ .... $\times$ .... $+ c$

$c = $ ....

$y = $ .... $x - $ ....   **(2 marks)**

.... **(2 marks)**

> Substitute the value of the gradient into $y = mx + c$. Then substitute the x-values and y-values given into your equation. Solve the equation to find c. Remember to write your completed equation at the end.

(c) gradient 4, passing through the point $(-2, 7)$

(d) gradient 4, passing through the point $(-1, -6)$.

.... **(2 marks)**

.... **(2 marks)**

6 Work out the equation of a straight line which passes through these points.

> **Guided**

(a) (3, 2) and (5, 6)

(b) $(-1, 5)$ and $(4, -10)$

$m = $ ....

$2 = $ .... $\times$ .... $+ c$

$c = $ ....

$y = $ .... $x - $ ....   **(3 marks)**

.... **(3 marks)**

7

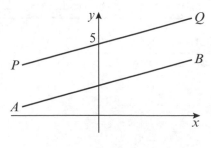

Here are two straight lines.
The equation of line AB is $y = 4x + 1$
Line AB is parallel to line PQ
Work out the equation of line PQ

> Find the gradient of AB.
> Find the value of the y-intercept for PQ.
> Use $y = mx + c$ to find the equation of a straight line.

.... **(2 marks)**

Real-life graphs

1

Guided

You can use this graph to change between miles and kilometres.

(a) Use the graph to change 60 miles into kilometres.

60 miles = ... kilometres   **(1 mark)**

> Draw a vertical line from 60 miles to the line and then draw a horizontal line across to the kilometres.

(b) The distance from Rome to Lyon is 660 miles.
The distance from Rome to Marseille is 950 kilometres.
Is Rome closer to Lyon or closer to Marseille?
You **must** show your working.

> The horizontal scale on the graph does not go up to 660 miles. Use your answer to part (a) to work out 660 miles in kilometres, then write a conclusion.

(3 marks)

2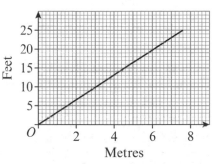

You can use this graph to change between feet and metres.

(a) Use the graph to change 15 feet into metres.

.............................m   **(1 mark)**

Amy and Sandeep are throwing a shot putt.

(b) Amy throws the shot putt 18 feet and Sandeep throws it 6 metres.
Who throws the shot putt the furthest? You **must** show your working.

(3 marks)

3 Lisa lays lawns in gardens of different areas.
She uses this graph to work out the cost of laying the lawn.

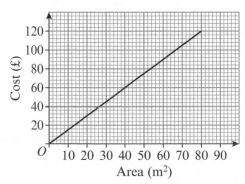

(a) She lays down 40 m² of lawn.
Use the graph to find the cost
of laying the lawn.

£............................   **(1 mark)**

(b) Lisa says, 'My price increases by £1.50
for every square metre.'
Is she correct?
Give a reason for your answer.

> The gradient of the graph gives the cost per m².

....

.... **(3 marks)**

Distance–time graphs

1 Becky cycled from her home to the shop. She went into the shop. She then cycled back home. Here is a distance–time graph for Becky's complete journey.

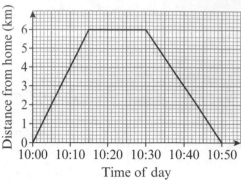

(a) What time did Becky start her journey?

Start time =

.... **(1 mark)**

(b) What is the distance from Becky's home to the shop?

> After how many kilometres does she stop?

......................................km **(1 mark)**

(c) How many minutes was Becky in the shop?

> This is where the graph is horizontal.

.... **(1 mark)**

(d) Work out Becky's average speed for her return journey.

speed = $\dfrac{\text{distance}}{\text{time}}$ = ... = ... km/h   **(2 marks)**

2 Gary left home at 1 pm to go for a walk. The distance–time graph represents part of Gary's journey.

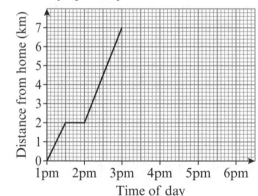

(a) Gary stopped for a break at 1.30 pm. Write down how many minutes Gary stopped for.

.... **(1 mark)**

(b) How far was Gary from home at 2 pm?

> Draw a line up from 2 pm and then across to the vertical axis.

.... **(1 mark)**

Gary had a rest at 3 pm for 1 hour. He then walked home at a steady speed. His walk home took him one and a half hours.

(c) Complete the distance–time graph.   **(2 marks)**

3 Nisha drives 40 km from her home to her grandparent's home. The journey took 1 hour. She spent 3 hours at their house, then drove home at a steady speed. Her journey home took 30 minutes.

(a) Draw a distance–time graph of Nisha's trip.

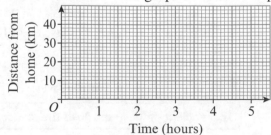

(2 marks)

(b) Work out Nisha's average speed for her return journey.

......................................km/h **(2 marks)**

Rates of change

1 Here are four flasks. Rachael fills each flask with water.
The graphs show the rate of change of the depth of the water in each flask as Rachael fills it.
Draw a line from each flask to the correct graph. One line has been drawn for you.

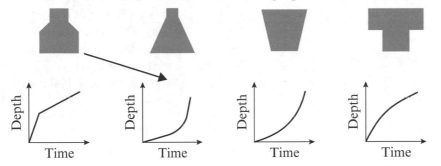

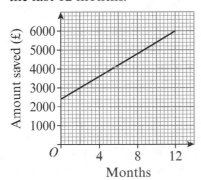

(2 marks)

2 The graph shows how much money there was in Dan's savings account over the last 12 months.

(a) How much money was there at the start?

£.. **(1 mark)**

(b) Work out the gradient of the line.

> Draw a triangle on the graph and use it to find the gradient.

.... **(2 marks)**

(c) Interpret the value of the gradient.

.... **(1 mark)**

3 Here is a velocity–time graph of a car.

> Draw a triangle on the graph and use it to find the gradient.

(a) What is the rate of change of velocity in the first 10 seconds?

$$\text{Gradient} = \frac{\text{distance up}}{\text{distance across}} = \frac{\text{..}}{\text{..}} = \text{..} \ \text{m/s}^2$$

(2 marks)

(b) Describe what is happening to the car between 10 seconds and 30 seconds.

> This is where the graph is horizontal.

.... **(1 mark)**

(c) What is the rate of change of velocity in the last 20 seconds?

.... m/s^2 **(2 marks)**

Expanding double brackets

1 Expand and simplify

> **Guided**

(a) $(x + 3)(x + 5)$

> 'Expand' means 'multiply out the brackets'.
> 'Simplify' means 'collect together like terms'.

$= x(x + 5) + 3(x + 5)$

> This is the 'one at a time' method. You can use any method you like to expand the brackets.
> The FOIL method is shown in question 2.

$= x^2 + \text{.................} + \text{.................} + \text{.................}$

$= x^2 + \text{.................} + \text{.................}$

(2 marks)

(b) $(x + 3)(x + 2)$

(c) $(x + 1)(x + 4)$

.... **(2 marks)**   .... **(2 marks)**

(d) $(x + 2)(x - 5)$

(e) $(x - 2)(x - 5)$

.... **(2 marks)**   .... **(2 marks)**

2 Expand and simplify

> **Guided**

(a) $(3x - 4)(5x - 1)$

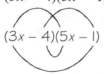

$(3x - 4)(5x - 1)$

> This is the 'FOIL' method. Multiply the First terms, then the Outer terms, then the Inner terms, then the Last terms.

$= 15x^2 - \text{...........................} - \text{...........................} + \text{...........................}$

$= 15x^2 - \text{...........................} + \text{...........................}$ **(2 marks)**

(b) $(8x - 3)(2x - 1)$

(c) $(7x - 5)(4x - 5)$

.... **(2 marks)**   .... **(2 marks)**

(d) $(x + 3)^2$

(e) $(2x - 5)^2$

.... **(2 marks)**   .... **(2 marks)**

3 In the diagram, ABCH is a rectangle, HCFG is a square and CDEF is a rectangle.

They are joined to make an L-shape.

$AH = (x + 3)\,\text{cm}$, $AB = (x + 2)\,\text{cm}$ and $CD = 4\,\text{cm}$

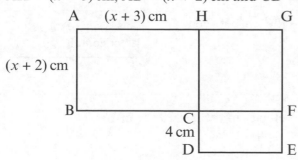

Show that the total area of L-shape, in cm², is $2x^2 + 13x + 18$ **(4 marks)**

Quadratic graphs

> **Guided**

1 (a) Complete the table of values for $y = x^2 - 2$

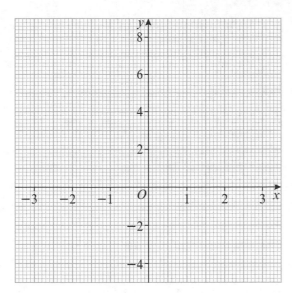

x	-3	-2	-1	0	1	2	3
y		2					7

> Substitute each value of x into the rule $y = x^2 - 2$ to find the value of y.

$x = -3: y = (-3 \times -3) - 2$

$= \ldots\ldots\ldots\ldots\ldots - 2$

$= \ldots\ldots\ldots\ldots\ldots$

$x = 1: y = (1 \times \ldots\ldots\ldots) - 2$

$= \ldots\ldots\ldots\ldots\ldots - 2$

$= \ldots\ldots\ldots\ldots\ldots$ **(2 marks)**

(b) On the grid draw the graph of $y = x^2 - 2$ **(2 marks)**

(c) Write the coordinates of the turning point.

> The turning point is the point where the direction of the curve changes.

$\ldots\ldots\ldots\ldots\ldots$ **(1 mark)**

(d) Use your graph to work out the value of y when $x = 2.5$ $\ldots\ldots$ **(1 mark)**

> **Guided**

2 (a) Complete the table of values for $y = x^2 - 4x + 3$

x	-1	0	1	2	3	4	5
y			0				8

> Substitute each value for x into the rule $y = x^2 - 4x + 3$ to find the value of y.

$x = -1: y = (-1 \times -1) - (4 \times -1) + 3 = \ldots\ldots\ldots\ldots$

$x = 3: y = (3 \times \ldots\ldots\ldots) - (4 \times \ldots\ldots\ldots) + 3$

$= \ldots\ldots\ldots\ldots$ **(2 marks)**

(b) On the grid draw the graph of $y = x^2 - 4x + 3$ **(2 marks)**

(c) Write the coordinates of the turning point. $\ldots\ldots\ldots\ldots\ldots$ **(1 mark)**

Using quadratic graphs

> **Guided**

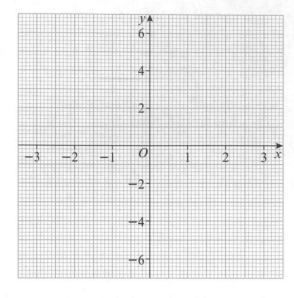

1 (a) Complete the table of values for
$y = x^2 - x - 4$

x	−2	−1	0	1	2	3
y			−4			2

$x = -2$: $y = (-2 \times -2) - (-2) - 4$

= ....

$x = 1$: $y = (1 \times 1) - (1) - 4$

= .... **(2 marks)**

(b) On the grid draw the graph of
$y = x^2 - x - 4$ **(2 marks)**

(c) Use your graph to write an estimate for
the minimum value of y

> The minimum value of y is the point where
> the direction of the curve changes.

.... **(1 mark)**

(d) Use your graph to write an estimate for
the solutions of $x^2 - x - 4 = 0$

.... **(2 marks)**

> **Guided**

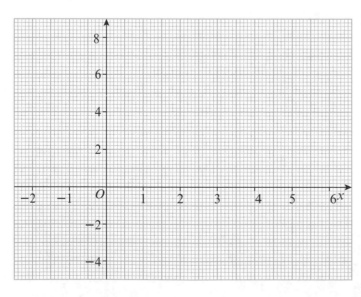

2 (a) Complete the table of values for
$y = 2 + 4x - x^2$

x	−1	0	1	2	3	4	5
y		2			5		

$x = 2$: $y = 2 + (4 \times 2) - (2 \times 2)$

= ....

$x = -1$: $y = 2 + (4 \times 1) - (-1 \times -1)$

= .... **(2 marks)**

(b) On the grid draw the graph of $y = 2 + 4x - x^2$ **(2 marks)**

(c) Write the coordinates of the turning point. .... **(1 mark)**

(d) Estimate the roots of the equation $2 + 4x - x^2 = 0$

> The 'roots' of the equation are the x-values
> where the graph intersects the x-axis.

.... **(2 marks)**

Factorising quadratics

1 Factorise

(a) $x^2 + 4x + 3$ $\boxed{\text{You need to find two numbers that multiply to make 3 and add up to 4.}}$

(..................) × (..................) = +3   (..................) × (..................) = +4

$x^2 + 4x + 3 = (x +$ ....$)(x +$ ....$)$ **(2 marks)**

(b) $x^2 + 11x + 10$

(..............) × (..............) = +10   (..............) + (..............) = +11

$x^2 + 11x + 10 = (x$ ....$)(x$ ....$)$ **(2 marks)**

(c) $x^2 + 6x + 5$ (d) $x^2 - 11x + 10$ (e) $x^2 - 12x + 20$

.... **(2 marks)** .... **(2 marks)** .... **(2 marks)**

(f) $x^2 - 9x + 14$

(..............) × (..............) = +14   (..............) + (..............) = −9

$x^2 - 9x + 14 = (x$ ....$)(x$ ....$)$ **(2 marks)**

2 Factorise

(a) $x^2 + 6x - 7$ (b) $x^2 + 4x - 5$ (c) $x^2 - 2x - 15$

.... **(2 marks)** .... **(2 marks)** .... **(2 marks)**

3 Factorise

(a) $x^2 - 13x + 22$ (b) $x^2 - 6x - 16$ (c) $x^2 - 14x + 40$

.... **(2 marks)** .... **(2 marks)** .... **(2 marks)**

4 Factorise

(a) $x^2 - 9$

$\boxed{\text{This is a difference of two squares. You can use the rule } a^2 - b^2 = (a + b)(a - b).}$

$a = x, b = 3$

$x^2 - 9 = (x +$ ....$)(x -$ ....$)$ **(2 marks)**

(b) $x^2 - 144$ (c) $x^2 - 81$ (d) $x^2 - 64$

.... **(2 marks)** .... **(2 marks)** .... **(2 marks)**

(e) $x^2 - 1$ (f) $x^2 - 169$

.... **(2 marks)** .... **(2 marks)**

Quadratic equations

1 Solve

(a) $x^2 - 3x = 0$

$x(x - \text{.....................}) = 0$

$x = 0$ or $x = \text{.....................}$ **(2 marks)**

> Find the values of x that make each factor equal to 0. The first factor is just x so one solution is $x = 0$.

(b) $x^2 + 5x = 0$

(c) $x^2 - 7x = 0$

.... **(2 marks)**　　.......................... **(2 marks)**

2 Solve

(a) $x^2 + 6x + 8 = 0$

> The first factor is $x + 2$, so the first solution is $x = -2$.

$(x + 2)(x + \text{.....................}) = 0$

$x = -2$ or $x = \text{.....................}$ **(3 marks)**

(b) $x^2 - 7x + 12 = 0$

(c) $x^2 + 9x + 20 = 0$

.... **(3 marks)**　　.......................... **(3 marks)**

(d) $x^2 + 8x + 7 = 0$

(e) $x^2 - 2x - 24 = 0$

.... **(3 marks)**　　.......................... **(3 marks)**

3 Solve

> Use the rule for the difference of two squares: $a^2 - b^2 = (a + b)(a - b)$.

(a) $x^2 - 4 = 0$

$(x + \text{.....................})(x - \text{.....................}) = 0$

$x = \text{.....................}$ or $x = \text{.....................}$ **(3 marks)**

(b) $x^2 - 25 = 0$

(c) $x^2 - 49 = 0$

.... **(3 marks)**　　.......................... **(3 marks)**

(d) $x^2 - 121 = 0$

(e) $x^2 - 9 = 0$

.... **(3 marks)**　　.......................... **(3 marks)**

4 Circle the equation with roots 2 and -7

$2x(x + 7) = 0$　　　$x^2 - 14 = 0$　　$(x - 2)(x - 7) = 0$　　$(x - 2)(x + 7) = 0$ **(1 mark)**

Cubic and reciprocal graphs

 Guided

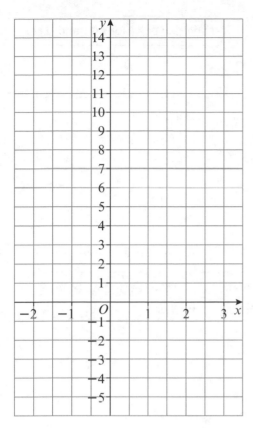

1 (a) Complete the table of values for $y = x^3 - 4x - 2$

x	−2	−1	0	1	2	3
y		1				13

> Substitute each value for x into the rule $y = x^3 - 4x - 2$ to find the value of y.

$x = -2$: $y = (-2 \times -2 \times -2) - (4 \times -2) - 2$

$= \dots\dots\dots\dots\dots$

$x = 1$: $y = (1 \times 1 \times 1) - (4 \times 1) - 2$

$= \dots\dots\dots\dots\dots\dots\dots$ **(2 marks)**

(b) On the grid draw the graph
     of $y = x^3 - 4x - 2$.   **(2 marks)**

(c) Estimate the value of x when $y = 5$

$\dots\dots\dots\dots\dots\dots\dots$ **(1 mark)**

(d) Use your graph to write down estimates of the solutions to the equation
     $x^3 - 4x - 2 = 0$ **(3 marks)**

2

A	B	C	D
	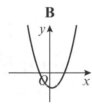		

Write down the letter of the graph which could have the equation

(a) $y = x^2 - x - 6$ (b) $y = x^3 - 3x + 5$

$\dots\dots\dots\dots$ **(1 mark)** $\dots\dots\dots\dots$ **(1 mark)**

(c) $y = \dfrac{1}{x}$ (d) $y = 6 - x - x^2$

$\dots\dots\dots\dots$ **(1 mark)** $\dots\dots\dots\dots$ **(1 mark)**

Simultaneous equations

1 Solve the simultaneous equations.

> Label the equations (1) and (2).

Guided

(a) $5x + y = 10$ (1)

       $x + 3y = 9$ (2)

   (1) × 3 gives $15x +$ .... $=$ .... (3)

   (3) − (2) gives   ....x $=$ ....

$x =$ ....

Substitute $x =$ .... in (1)

$5 ×$ .... $+ y = 10$

$y =$ ....

$x =$ ...., $y =$ .... **(3 marks)**

(b) $3x + 2y = 11$

       $2x - 5y = 20$

> Multiply equation (1) by 5 and equation (2) by 2.
> Then add the new equations.

$x =$ ...., $y =$ .... **(4 marks)**

2 By drawing two suitable straight lines on the coordinate grid below, solve the simultaneous equations.

(a) $x + y = 5$ (b) $2x + y = 5$

      $y = 3x + 1$ $x + y = 3$

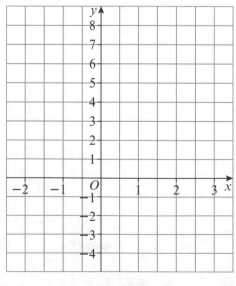
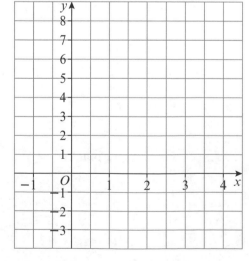

(a) $x =$ ....,   (b) $x =$ ....,

    $y =$ .... **(3 marks)** $y =$ .... **(3 marks)**

> For each equation choose three x-values and then find the corresponding y-values.
> Plot these points and draw a straight line through the three points.
> The solution to the simultaneous equations is the point where the lines cross.

Rearranging formulae

1 A straight line has equation $3y = 4x + 3$. The point P lies on the straight line. P has y-coordinate of 5. Work out the x-coordinate of P.

> Substitute $y = 5$ into $3y = 4x + 3$ and solve the equation to find x.

$3 \times \text{........................} = 4x + 3$

$\text{........................} - \text{........................} = 4x$

$x = \text{........................} \div \text{........................} = \text{........................}$ **(3 marks)**

2 Mandy wants to work out the time, t seconds, taken for a ball to reach the ground by using the formula $v = u + at$. She knows that $v = 30$, $u = 5$ and $a = 10$. Work out the time taken for the ball to reach the ground.

$t = \text{........................}$ **(3 marks)**

3 Make the letter in the brackets the subject of each formula.

(a) $v = u + 10t$ (t) (b) $m = 6n + 19$ (n)

$v - \text{........................} = 10t$ $(\div 10)$

$\text{........................} = t$

$t = \text{........................}$ **(2 marks)** $n = \text{..}$ **(2 marks)**

(c) $d = ut + at^2$ (u) (d) $P = A - 6D$ (D)

$u = \text{..}$ **(2 marks)** $D = \text{..}$ **(2 marks)**

4 Make the letter in the brackets the subject of each formula.

(a) $s = \dfrac{d}{t}$ (t) (b) $d = \sqrt{\dfrac{5h}{4}}$ (h) > First square both sides.

$t = \text{..}$ **(2 marks)** $h = \text{..}$ **(2 marks)**

(c) $s = \dfrac{1}{2}(u + v)t$ (t) (d) $v^2 = u^2 + 2as$ (s)

$t = \text{..}$ **(2 marks)** $s = \text{..}$ **(2 marks)**

5 Make the letter in the brackets the subject of each formula.

(a) $P = h(2 + n)$ (n) (b) $t = 3(1 - 2x)$ (x) > Multiply out the brackets first.

$n = \text{..}$ **(3 marks)** $x = \text{..}$ **(3 marks)**

Using algebra

1 Tom plants some rhubarb seeds in his allotment.
He plants seven rows of these seeds. In each row there are x seeds.
A few months later he finds insects have eaten nine seeds.
After this he has 96 rhubarb plants left in total.

> **Guided**

(a) Write an equation using this information.

$7 \times$ .... $-$ .... $=$ .... **(1 mark)**

(b) Work out the value of x

> Solve the equation.

.... **(2 marks)**

2 The diagram shows a triangular playground. Form an equation, then simplify and solve it.

> **Guided**

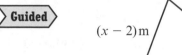

$(x - 2)\,\text{m}$ $(x + 5)\,\text{m}$

$(2x - 9)\,\text{m}$

The total perimeter of the playground is 54 m.
Work out the value of x

$(x - 2) +$ .... $+$ .... $= 54$

.... **(3 marks)**

3 A square has side length $2x$ cm.
An equilateral triangle has side length $(2x + 4)$ cm.
The perimeter of the square is equal to the perimeter of the equilateral triangle.
Work out the value of x

$x =$ .... **(3 marks)**

4 The diagram shows two rectangular gardens. All the measurements are in metres.

$3x - 2$ | A

$2x + 3$

$2x + 1$ | B

$2x + 6$

Both gardens have the same perimeter. Work out the width and length of garden A.

.... **(4 marks)**

Identities

1 p is a prime number and q is an odd number.

Is the expression $p^2 + 3q$ always odd, always even or could it be either odd or even?

Tick a box.

Always odd ☐   Always even ☐   Could be odd or even ☐

Give examples to justify your answer.

....

.... **(1 mark)**

2 h and w are **both** prime numbers.

Work out values for h and w so that $h = w^2 + 6$

> **Guided**

....

.... **(2 marks)**

3 x, y and z are all odd numbers.

Which of these expressions has a value that is always even?

Circle your answer.

$\quad\quad\quad x + y + z \quad\quad\quad xyz \quad\quad\quad x^2 + yz \quad\quad\quad xy + 2z$ **(1 mark)**

4 Show that

> **Guided**

(a) $(x - 2)^2 \equiv x^2 - 4x + 4$ | '$\equiv$' is the identity symbol. |

$(x - 2)^2 = (x - 2)(x - 2) = x^2 - \text{............} - \text{............} + \text{............} = \text{....................}$

(2 marks)

(b) Hence, or otherwise, show that | 'Hence' means that you can use your
$(x + 2)^2 - (x - 2)^2 \equiv 8x$ answer to part (a) to help answer part (b). |

$(x + 2)^2 - (x - 2)^2 = (x + 2)(x + 2) - (\text{....................})(\text{....................})$

$= x^2 + \text{............} + \text{............} + \text{............} - (\text{................................})$

$= \text{................................} - \text{................................}$

$= 8x$ | Be careful with the signs. | **(3 marks)**

5 Circle the expression that is equivalent to $y^2 - 6y - 16$

$\quad\quad (y - 4)^2 \quad\quad (y - 2)(y + 8) \quad\quad (y + 2)(y - 8) \quad\quad (y + 4)(y - 4)$ **(1 mark)**

Problem-solving practice 1

1   Tom cleaned his swimming pool. He hired a cleaning machine to do this job.
The cost of hiring the cleaning machine was £35.50 for the first day and then
£18.25 for each extra day.
Tom's total cost of hiring the machine was £163.25
For how many days did Tom hire the machine?

.... days   **(3 marks)**

2   The coordinates of three vertices of a square are (2, 1), (2, 5) and (6, 5).

Write the coordinates of the

(a) fourth vertex   (b) centre of the square.

.... **(1 mark)**   .... **(2 marks)**

3   A gardener charges £15 for each hour he works at a job plus £25. The cost, in £,
of the job can be worked out using the formula
     Cost = number of hours worked × 15 + 25

(a) The gardener works seven hours. Work out the total cost.

£..................................... **(2 marks)**

(b) He charges £115 for one job. How many hours did he work?

.... hours   **(3 marks)**

4   Dan and Jay are measuring the distance, s, when a ball is rolled.
They use the formula $s = ut + 5t^2$ where initial speed, $u = 4$ and time, $t = 2$
Dan works out the value of s to be 28 and Jay works out the value of s to be 108
Who is correct? Give a reason for your answer.

.... **(3 marks)**

5   Triangle ABC is isosceles with AB = BC

$5(x - 1)$ cm $2x + 4$ cm

B _____ C
    $3(x + 2)$ cm

Work out the lengths of the sides of the triangle.

.... **(4 marks)**

Problem-solving practice 2

6 The diagram shows a square.
Work out the area of the square.

$(30 + x)$

$(45 - 2x)$

.... **(3 marks)**

7 ABC is a triangle. Work out the size of the smallest angle.

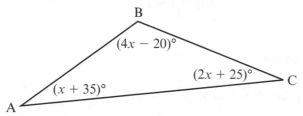

B

$(4x - 20)°$

$(2x + 25)°$ C

$(x + 35)°$

A

....................................° **(3 marks)**

8 (a) Write the equation of a straight line that is parallel to $y = 5x + 4$

.... **(1 mark)**

(b) Work out the equation of the straight line that is parallel to $y = 3x + 5$
and which passes through $(4, 7)$.

.... **(3 marks)**

9 Here are the first five terms of a linear sequence.

2   5   8   11   14

Jo says that 234 is a term in this sequence.
Is she correct?
Show working to justify your answer.

.... **(3 marks)**

10 Here are two number machines A and B.

Input   Output

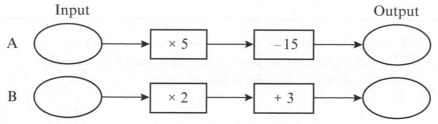

A   ( ) → × 5 → − 15 → ( )

B   ( ) → × 2 → + 3 → ( )

Both machines have the same input.

Work out the input that makes the output of A **four** times the output of B.   **(4 marks)**

Percentages

> 'Per cent' means 'out of 100'.

1 Work out

(a) 9% of 50

$\dfrac{9}{100} \times$ .... = ...

(2 marks)

(b) 4% of 275

.... **(2 marks)**

2 Work out the following as percentages.

(a) 35 out of 56

$\dfrac{35}{56} \times$ .... = ...

(2 marks)

(b) 15 out of 75

.........................% **(2 marks)**

3 A ticket for the theatre costs £96 plus a booking charge of 8%. Work out

(a) the amount of the booking charge

£......................... **(2 marks)**

(b) the total cost of a theatre ticket.

£......................... **(1 mark)**

4 Suki invited 128 people to a Christmas party. 48 people were adults.
Work out 48 as a percentage of 128

.........................% **(2 marks)**

5 A gym has 450 members.
28% have 'off-peak' membership, 16% have 'family' membership.
The rest have 'full' membership.

Work out the number of 'full' members. **(3 marks)**

6 The table gives information about boys who choose French or German and girls who choose French or German at school.

	Number of boys	Number of girls
French	84	56
German	54	126

(a) Work out the percentage of girls who chose French.

> Start by finding the total number of girls.

.........................% **(3 marks)**

(b) 30% of the students who chose French passed their exam.
60% of the students who chose German passed their exam.
Show that 47% of all students passed their exam.

(4 marks)

Fractions, decimals and percentages

1 Write these percentages as fractions in their simplest form.

(a) 24%

Simplify your fraction if possible.

$$24\% = \frac{24}{100} = \text{..................}$$
....

(1 mark)

(b) 64%

(c) 85%

.... **(1 mark)**

.... **(1 mark)**

2 Write these numbers in order, starting with the smallest.

(a) 0.62 $\frac{3}{10}$ 61%

(b) 33% 0.32 $\frac{7}{20}$

(c) 0.38 37% $\frac{2}{5}$

.... **(2 marks)**

.... **(2 marks)**

.... **(2 marks)**

3 John earns £2300 per month. He spends 15% of his salary as rent and $\frac{3}{5}$ of his salary on bills.
Work out how much John has left after he has paid his rent and bills.

You will need to use problem-solving skills throughout your exam – **be prepared!**

Plan your strategy before you start. You'll save time if you convert $\frac{3}{5}$ into a percentage.

$\frac{3}{5}$ = ... %

15% + ... % = ... %

100% − .... % = ... %

.... % of £2300 = £..

(3 marks)

4 There are 120 students in Year 11.
$\frac{1}{6}$ of the students travel to school by bike.
35% of the students travel to school by car.
The rest of the students walk to school. How many students walk to school?

.... **(3 marks)**

5 At the cinema, 64% of the audience are children.
$\frac{2}{3}$ of the remaining audience are women.
There are 30 men in the audience.

How many people are at the cinema?

.... **(3 marks)**

Percentage change 1

1 Circle the multiplier you would use to calculate a percentage increase of 8%.

> Work out 100% + 8% then change this percentage to a decimal.

 Guided

1.18   1.8   0.92   1.08   **(1 mark)**

 Guided

2 (a) Decrease 78 by 4%. Start by finding 4% of 78.   (b) Increase 126 by 4%.   You can use a multiplier.

$$\frac{4}{100} \times \text{...............} = \text{...............}$$

78 − .... = ... **(2 marks)**   .... **(2 marks)**

(c) Decrease 96 by 12%.   (d) Increase 242 by 14%.

.... **(2 marks)**   .... **(2 marks)**

3 A shop sells a mobile phone for £135. A discount of 6% is given. Work out the price of the mobile phone after the discount.

£........................ **(2 marks)**

 Guided

4 Work out the percentage change of these price changes.

	Original price (£)	New price (£)	Percentage change
(a)	640	512	$\dfrac{640 - \text{...............}}{640} \times 100 = \text{...............}$%
(b)	160	208	...............%
(c)	1560	2106	...............%
(d)	2750	2475	...............%

(12 marks)

5 Chloe is collecting reward points in an online video game.

She collected 5100 points last month and 3672 points this month.
Work out the percentage decrease in the number of points she collected.

........................% **(3 marks)**

6 Niamh and Owen received the same percentage pay rise in 2015.
In 2014 Niamh earned £24 500 per year.
In 2015, after the pay rise, she earned £25 970
In 2014 Owen earned £22 000
Work out Owen's salary in 2015.

£........................ **(4 marks)**

Percentage change 2

7 Aaron is comparing the additional charges for two airlines.

Tricky-jet	**Kelly-air**
Credit card charge: 3%	Credit card charge: 5%
Booking fee: £5	Booking fee: £2

You will need to use problem-solving skills throughout your exam – **be prepared!**

A ticket is advertised as costing £90 from both airlines. Work out which airline is cheaper after the additional charges are applied.

Tricky-jet

$\dfrac{3}{..............}$ × .... = ...

£.................... + £.................... + £.................... = £....................

Kelly-air

$\dfrac{5}{..............}$ × .... = ...

£.................... + £.................... + £.................... = £....................

.... is cheaper.   **(5 marks)**

8 Zac wants to buy some concrete posts. He finds two companies on the internet.

Postland	**C & R**
10 posts for £10.50 each and receive a 15% discount	5 posts for £37.75 plus 20% VAT

Zac needs to buy ten concrete posts and wants the cheapest option.
Which of the two companies should Zac buy the concrete posts from?

Remember to show all your working and write a conclusion.

.... **(5 marks)**

9 Sandeep wants to buy a pair of trainers. He finds that two online shops sell the trainers he wants.

Footworld	**Sportish**
£42.50 for a pair	£30.90 for a pair plus
Online discount of 22%	VAT at 20%

Sandeep wants to pay the lowest price. Which shop should Sandeep buy his trainers from?

.... **(5 marks)**

Ratio 1

1 Write these ratios in their simplest form.

(a) 45 : 30

> Divide both parts of the ratio by the same number.

45 : 30

÷ .... (............) ÷ ....

.... : ....

(2 marks)

(b) 54 : 16

.... **(2 marks)**

(c) 56 : 64

.... **(2 marks)**

2 (a) Divide £50 in the ratio 2 : 3

Total parts = ... + ... = ...

1 part = 50 ÷ .... = ...

2 parts = ... × .... = ...

3 parts = ... × .... = ...

(2 marks)

(b) Divide £750 in the ratio 2 : 5 : 8

.... **(3 marks)**

3 Sandeep is going to make a pizza. He uses cheese, peppers and dough in the ratio 2 : 3 : 7
He uses 56 g of dough. Work out the number of grams of cheese and the number of grams of peppers he uses.

> Seven parts of the ratio represents 56 g. Work out how much one part of the ratio represents.

cheese .... g

pepper .... g **(3 marks)**

4 Anjali, Paul and Faye are travelling in a car from Wolverhampton to London.
They share the driving so that the distances driven are in the ratio 3 : 4 : 5 respectively.
Anjali drives the least distance of 36 miles.
Work out the distances Paul and Faye each drive.

Paul .... miles

Faye .... miles **(3 marks)**

5 Amish, Benji and Cary save some money in the ratio 3 : 4 : 9
Cary saved £120 more than Benji.

(a) Show that Amish saved £72

(b) Show that the total amount of money saved was £384

(2 marks)

(1 mark)

Ratio 2

6 Solder is made from lead and tin. The ratio of lead to tin is 2 : 3

 Guided

(a) Kyle made 70 g of solder. Work out the mass of the lead used.

Total parts = ... + ... = ...

1 part = 70 ÷ .... = ... = ...

Lead = 2 parts = 2 × .... = ... = ...   **(2 marks)**

(b) He then uses 16 g of lead to make some more solder.
Work out the mass of solder he made.

.... g **(2 marks)**

7 Gabby and Harry shared some money in the ratio 3 : 8
Harry received £2000 more than Gabby.
How much money did they share?

Guided

PROBLEM SOLVED!

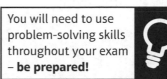
You will need to use problem-solving skills throughout your exam – **be prepared!**

8 − 3 = 5 parts

5 parts = £2000

1 part = £2000 ÷ 5 = ...

11 parts = 11 × .... − ....

In total they shared £.............

Harry received £2000 more than Gabby and 8 − 3 = 5 so five parts of the ratio represents £2000.

(3 marks)

8 Asha uses an old recipe to make some cakes.
The ratio of the masses of the flour, margarine and sugar needed for the recipe is 5 : 4 : 3
Asha has the following amounts of each ingredient.

1825 g of flour
700 g of margarine
250 g of sugar

Each cake needs 48 g of the combined ingredients. Show that the maximum number of cakes she can make is 20.

(3 marks)

9 White paint costs £2.60 per litre. Red paint costs £3.80 per litre.
They are mixed in the ratio white : red = 5 : 3
Work out the cost of 15 litres of the mixture.

....

.... **(4 marks)**

Metric units

Guided

1 Change

(a) 45 mm to cm

> You need to remember the metric conversions.

45 ÷ .... = ... *cm*   **(1 mark)**

(b) 72 cm to mm

72 × .... = ... mm   **(1 mark)**

(c) 3.5 km to m

(d) 5.3 kg to g   > 'Kilo' means 'one thousand'.

.....................m   **(1 mark)**   ....g   **(1 mark)**

(e) 4.3 litres to ml

(f) 480 mg to g

.....................ml   **(1 mark)**   ....g   **(1 mark)**

2 Change

(a) 15 cm to mm

(b) 28 mm to cm

.....................mm   **(1 mark)**   ....cm   **(1 mark)**

(c) 1800 g to kg

(d) 2800 m to km

.....................kg   **(1 mark)**   ....km   **(1 mark)**

3 How many mm are there 15.2 metres?

Circle your answer.

0.0152   152   15 200   152 000   **(1 mark)**

4 How many 125 ml cups can be filled from a bottle holding 2 litres of squash?

> Convert 2 litres into ml.

.... **(2 marks)**

5 How many 75 mm pieces of wood can be cut from a piece of wood of length 6.2 m?

.... **(2 marks)**

6 A bookshelf is 1 m wide.
Joe wants to place a set of books across the shelf.
He has 20 books. Each book is 5.2 cm wide.
Will he have enough space for all the books?

> Don't just answer 'yes' or 'no'. You need to show your working then write a conclusion.

.... **(3 marks)**

Reverse percentages

1 In a sale all prices are reduced by 30%.
Andy buys a shirt on sale for £42
Work out the original price of the shirt.

> First work out the multiplier for a 30% decrease.

100% − 30% = ...%

$\dfrac{............}{100}$ = ...

£42 ÷ .... = £............

(3 marks)

2 Brinder receives a pay rise of 6%.
After the pay rise, Brinder earns a salary of £35 245.
Work out Brinder's salary before the pay rise.

> First work out the multiplier for a 6% increase.

100% + ... = ...%

$\dfrac{............}{100}$ = ...

£35 245 ÷ .... = £............

(3 marks)

3 Kam bought a new car. The car depreciates by 15% each year.
After one year the car was worth £28 560
Work out the price of the car when it was new.

> Check that your answer makes sense. The original price of the car should be **greater** than £28 560.

£.. **(3 marks)**

4 In a sale, the original price of a phone is reduced by one-quarter.
The sale price of the phone is £368.97

Work out the original price of the phone.

(3 marks)

5 Alison and Nav invested some money in the stock market in 2014.

This table shows the value of their investments in 2015.

Who invested the most money originally?
You **must** show your working.

> You will need to use problem-solving skills throughout your exam – **be prepared!**

....

	Value in 2015	Percentage increase since original investment
Alison	£1848	12%
Nav	£1764	5%

(4 marks)

Growth and decay

1 Raj invests £12 000 for four years at 10% per annum compound interest.
Work out the value of the investment at the end of four years.

> First work out the multiplier for a 10% increase.

100% + 10% = ...%

$\dfrac{.............}{100}$ = ...

£12 000 × (.............)$^{.....}$ = £.............

(2 marks)

2 Neil invests £5800 at a compound interest rate of 6% per annum.
At the end of n complete years the investment has grown to £6907.89
Work out the value of n

> Choose some values of n and work out the amount of investment after n years.

n = ... **(2 marks)**

3 Omar invested £1500 at a fixed compound interest rate of 3.5% per annum.
Circle the amount he will have after eight years.

1500×1.35^8 $1500 \times 1.08^{3.5}$ 1500×0.035^8 1500×1.035^8 **(1 mark)**

4 (a) Chris bought a lorry that had a value of £24 000
Each year the value of the lorry depreciates by 15%.
Work out the value of the lorry at the end of four years.

> First work out the multiplier.

100% − ....% = ...%

$\dfrac{.............}{100}$ = ...

£24 000 × (.............)$^{.....}$ = £.............

(2 marks)

(b) Brian bought a new car for £12 000. Each year the value of the car depreciates by 12%. Work out the value of the car at the end of five years.

£............. **(2 marks)**

5 Daljit invests £1500 on 1 January 2010 at a compound interest rate of r% per annum. The value, £V, of this investment after n years is given by the formula $V = 1500 \times (1.065)^n$

(a) Work out the value of r

> Work out what percentage would give a multiplier of 1.065.

r = ... **(1 mark)**

(b) Work out the value of Daljit's investment after 10 years.

£...................................... **(2 marks)**

Speed

1 Anjali runs 400 m in 44.7 seconds.

Work out Anjali's average speed.

Give your answer to 1 decimal place.

Speed = distance ÷ time

Speed = ... ÷ .... = ...m/s

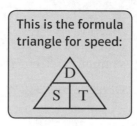

This is the formula triangle for speed:

D
S | T

(2 marks)

2 A car travels 80 km in 1 hour and 45 minutes.

Circle the calculation that will give the average speed of the car in km/h.

$$\frac{80}{1.45} \qquad \frac{80}{1.045} \qquad \frac{80}{1.75} \qquad \frac{80}{1.075}$$

(1 mark)

3 Sandeep drives 344 km at an average speed of 80 km/h.
Work out the time taken for Sandeep's journey.

Give your answer in hours and minutes.
0.3 hours = 0.3 x 60 minutes

.... hours .... minutes   **(3 marks)**

4 Selma drives for four hours. Her average speed is 60 km/h.
Work out the total distance she travels.

Distance = ... × ....

Always write down the formula.

Distance = ... × .... = ...km   **(2 marks)**

5 Pavan is driving in France. The legal speed limit on French motorways is
130 km/h. He travels from one junction to another in 15 minutes and he
covers a distance of 35 km. Show that he has broken the speed limit.

(3 marks)

6 Jane travelled 50 km in 1 hour 15 minutes.
Karen travelled 80 km in 2 hours and 45 minutes.
Who had the lower average speed? You **must** show your working.

(3 marks)

7 At a school's sports day the 100 m race was won in 14.82 seconds and the 200 m
was won in 29.78 seconds. Which race was won with a faster average speed?
You **must** show your working.

(3 marks)

Density

1 What is the density of a piece of wood that has a mass of 17.5 g and a volume of 20 cm³?

Density = mass ÷ volume

Density = ... ÷ .... = ... g/cm³

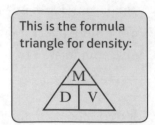

This is the formula triangle for density:

(2 marks)

2 Len has a silver ring which has a volume of 14 cm³. The density of silver is 10.5 g/cm³. Work out the mass of the silver ring.

Mass = ... × ....

Always write down the formula.

Mass = ... × .... = ... g

(2 marks)

3 Copper has a density of 8.96 g/cm³.
A piece of copper pipe has a mass of 5 kg.
Circle the calculation that will give the volume of the copper pipe in cm³.

$$\frac{5}{0.0896} \qquad \frac{5000}{8.96} \qquad \frac{8.96}{5} \qquad \frac{8960}{5}$$

(1 mark)

4 This solid cuboid is made of plastic.
The plastic has a density of 0.9 g/cm³.
Work out the mass of the cuboid.

8 cm
10 cm
6 cm

.... g **(4 marks)**

5 The diagram shows a solid triangular prism.
The prism is made of iron.
Iron has a density of 7.87 g/cm³.
Work out the mass of the prism.

8 cm
12 cm
15 cm

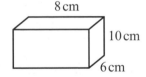

Remember, the volume of a prism is given by:
Volume = Area of cross-section × Length.

.... g **(4 marks)**

6 Gavin weighed some metal beads. They had a mass of 950 g.
The volume of the beads was 96 cm³.
Gavin worked out the density and claimed that the metal was gold.
Use the information in the table to work out whether Gavin is correct. You **must** show your working.

Metal	Density g/cm³
Gold	19.3
Copper	8.6
Bronze	9.9

(3 marks)

Other compound measures

1 A safe exerts a force of 600 N on the floor. The area of the base on the floor is 1.5 m². Work out the pressure exerted on the floor.

> This is the formula triangle for pressure:

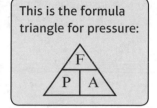

Pressure = force ÷ area

Pressure = ... ÷ .... = ... N/m²

> The force is in N and the area is in m², so the units of pressure will be N/m².

(2 marks)

2 Ray exerts a force of 900 N on the ground. His feet have an area of 0.024 square metres each. Work out the pressure he exerts on the ground.

> How many feet does Ray have?

.... N/m² **(3 marks)**

3 The pressure between a car's four tyres and the road is 400 000 N/m².
The car exerts a force of 10 000 N on the road.
Work out the area of contact between each tyre and the road.

Total area = ... ÷ ....

> Always write down the formula.

Area for 1 tyre = ... ÷ .... = ... m² **(3 marks)**

4 A cube of side 55 cm exerts a force of 400 N on a flat surface.
Work out the pressure in N/m². **(2 marks)**

5 An overflow pan at a factory can be modelled as a cuboid.

> You will need to use problem-solving skills throughout your exam – **be prepared!**

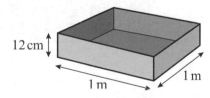

The pan is half-full of water.
The water flows out of the pan at an average rate of 250 millilitres per second.
Show that the pan will be completely empty after four minutes.

> 1 cm³ = 1 ml. Remember to convert metres to cm before calculating the volume of the cuboid.

(5 marks)

Proportion

1 David bought 4 kg of apples from the supermarket for £1.60
What is the cost of 7 kg of apples?

Guided

> Find the cost of 1 kg.

4 kg costs 160p

1 kg costs .... ÷ .... = ...

7 kg costs .... × .... = ... **(2 marks)**

2 A fabric shop sells material by the metre.
Andy bought 3 m of material for £2.25. What is the cost of 11 m of the same material?

£........................ **(2 marks)**

3 All tickets for the theatre are the same price.
Ben and Liam pay £75.90 altogether for some tickets.
Ben pays £31.05 for 9 tickets.

How many tickets does Liam buy? **(4 marks)**

4 An amount of money is divided equally among eight children. Each child receives £24
If the same amount of money was divided equally among 12 children, how much
would each child receive?

£........................ **(2 marks)**

5 A 200 g box of sweets costs £4.80
A 175 g box of sweets costs £4.50

Which size box is better value for money?

> Show all your working and then write a conclusion.

.... **(3 marks)**

6 Ten people take eight days to build a wall.
How long will it take four people to do the same job?

Guided

> Work out how long it will take one person to build the wall.

10 people work 8 days

1 person works .... × .... = ...
4 people work .... ÷ .... = ... days **(2 marks)**

7 A school building can be decorated by 12 people working 8 hours a day for
5 days. Mike wants to know how long it would take 10 people working
6 hours a day.

.... **(2 marks)**

Proportion and graphs

1 The force, F Newtons, on a mass is directly proportional to the acceleration, a m/s², of the mass. When $a = 25$, $F = 650$. Work out the value of F when $a = 45$ m/s².

Guided

Direct proportion, so $F = ka$ (where k is a constant)

When $a = 25$, $F = 650$ so $650 = k \times 25$

$k = 650 \div 25 = \ldots\ldots\ldots\ldots\ldots$

When $a = 45$ so $F = k \times 45 = \ldots\ldots\ldots\ldots\ldots \times 45 = \ldots\ldots\ldots\ldots\ldots$ Newtons **(2 marks)**

2 The resistance R ohms of a wire is inversely proportional to the cross-sectional area A cm² of a wire. When $A = 0.1$, $R = 30$. Work out the value of R when $A = 0.4$

Guided

Inverse proportion, so $R = \dfrac{k}{A}$ (where k is a constant)

When $A = 0.1$, $R = 30$ so $30 = \dfrac{k}{0.1}$

$k = 30 \times \ldots\ldots\ldots\ldots\ldots = \ldots\ldots\ldots\ldots\ldots$

When $A = 0.4$ $R = \dfrac{k}{0.4} = \dfrac{\ldots\ldots}{0.4} = \ldots\ldots\ldots\ldots\ldots$ ohms **(2 marks)**

3 x and y are inversely proportional. Circle the equation that could describe the relationship between x and y.

$x = 2y$ $\qquad$ $x = 3\sqrt{y}$ $\qquad$ $x = \dfrac{1}{2y}$ $\qquad$ $x = \dfrac{y}{5}$ **(1 mark)**

4 P is directly proportional to Q.
When $P = 12$, $Q = 0.25$

(a) Work out the value of P when
$Q = 0.4$ **(2 marks)**

(b) Work out the value of Q when
$P = 144$ **(2 marks)**

(c) On the axes, sketch the graph of
P against Q **(1 mark)**

5 W is inversely proportional to V.
When $W = 100$, $V = 32$

(a) Work out the value of V when
$W = 16$ **(2 marks)**

(b) Work out the value of W when
$V = 1280$ **(2 marks)**

(c) On the axes, sketch the graph of
V against W **(1 mark)**

Had a go ☐   Nearly there ☐   Nailed it! ☐

Problem-solving practice 1

1 Julie got 41 out of 50 marks in a mathematics test.
She got 50 out of 60 marks in a statistics test.
In which test did Julie get the higher percentage mark?
You **must** show your working.

(3 marks)

2 Karen wants to buy a game for her new console.
She finds that two online shops sell the game she wants.

Nile	**T-bay**
Game costs £35.50	Game costs £30.90 + VAT
Online discount 16%	VAT 20%
Delivery charge £2.75	No delivery charge

Karen wants to pay the lowest price.
Which shop should Karen buy her game from? You **must** show your working.

(5 marks)

3 58 boys, 32 girls and 190 adults are members of a Judo club.
70 more children join the club.
The number of girls is now 16% of the total number of members.
How many of the 70 children were boys?

.... **(4 marks)**

4 Avtar has a full 900 ml bottle of patio sealer. He is going to mix some of the patio sealer with water. Here is the information on the label of the bottle.

> Patio sealer (900 ml)
>
> Mix $\frac{1}{5}$ of the patio sealer with 5400 ml of water

Avtar is going to use 900 ml of water. How many millilitres of patio sealer should Avtar use? You **must** show your working.

(4 marks)

Problem-solving practice 2

5　Bird food is sold in 5 kg sacks for £8.99 and 13 kg sacks for £20.95
Which sack of bird food is the better value for money?
You **must** show your working.

(3 marks)

6　The ingredients for making an apple and almond crumble for four people are shown.
The ingredients that Rachel has are also shown.

Ingredients for **four** people	Rachel's ingredients
80 g flour	1.2 kg flour
60 g almonds	200 g almonds
90 g brown sugar	500 g brown sugar
60 g butter	250 g butter
4 apples	16 apples

She wants to make an apple and almond crumble for 15 people.
Does she have enough ingredients?
You **must** show your working.

(3 marks)

7　Taran employs eight people to plaster the walls of a building in six days.
He realises that he needs to plaster the walls in just four days.
Taran says that he needs three more people working at the same rate to
plaster the walls in four days.
Is he correct? You **must** show your working.

(2 marks)

8　(a)　Kim wants to save a deposit for a house. His target is to save £17 500 in four years.
He invests £14 000 in an ISA for four years at 6% per annum compound interest.
Does he have enough money for his deposit? You **must** show your working.

(3 marks)

　　(b)　Kim decides to wait until his original investment of £14 000 has doubled.
How many years will it take?
Show working to justify your answer.

(2 marks)

Symmetry

1 These shapes have lines of symmetry.
Draw the lines of symmetry as indicated.

(a) exactly **one** line of symmetry

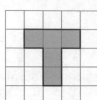

(1 mark)

(b) exactly **two** lines of symmetry

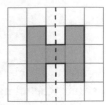

(1 mark)

2 Here are four shapes.

 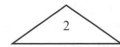

Write the number of a shape that has

(a) **no** lines of
symmetry

.... **(1 mark)**

(b) exactly **one** line
of symmetry

.... **(1 mark)**

(c) exactly **two** lines
of symmetry.

.... **(1 mark)**

3 Here is a regular hexagon.

(a) What is the order of rotational symmetry of the hexagon?

.... **(1 mark)**

(b) Draw a line of symmetry on the hexagon. **(1 mark)**

4 On each diagram, shade one square so that the shape has

(a) exactly **one** line of symmetry

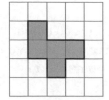

(1 mark)

(b) rotational symmetry of order two

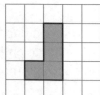

(1 mark)

(c) exactly **one** line of symmetry

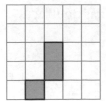

(1 mark)

(d) rotational symmetry of order two.

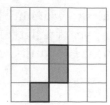

(1 mark)

Quadrilaterals

1 Write the mathematical name of each of these quadrilaterals.

Guided

(a)

....
Rectangle
(1 mark)

(b)

....
(1 mark)

(c)

....
(1 mark)

(d)

....
(1 mark)

(e)

....
(1 mark)

(f)

....
(1 mark)

2 Draw a quadrilateral with these properties.

(a) one pair of parallel sides

> Use arrows to show parallel lines, and dashes to show lines of equal length.

(1 mark)

(b) two pairs of parallel sides, opposite sides equal and opposite angles equal

(1 mark)

(c) two pairs of adjacent sides equal, one pair of opposite angles equal and diagonals cross at 90°

(1 mark)

3 This quadrilateral is drawn accurately.

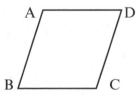

Which of these is the special name for quadrilateral ABCD?
Circle your answer.

    Kite   Rhombus   Trapezium   Rectangle   **(1 mark)**

Angles 1

 1 Draw lines from the name of the angle to the correct diagram.

acute   obtuse   reflex

(3 marks)

 2

(a) What type of angle is x?

....

(1 mark)

(b) Give a reason for your answer.

....

(1 mark)

 3 (a) What type of angle is a?

....

(1 mark)

(b) Give a reason for your answer.

....

(1 mark)

 4

Guided

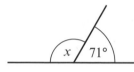

(a) Work out the size of the angle marked x

$x =$ .... $- 71° =$ ....° **(1 mark)**

(b) Give a reason for your answer.

Angles on a straight line add up to ....° **(1 mark)**

 5

Guided

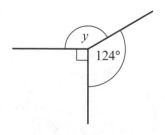

A right angle ⌐ is 90°

(a) Work out the size of the angle marked y

$y =$ ....° **(1 mark)**

(b) Give a reason for your answer.

Angles around a point add up to ....° **(1 mark)**

Angles 2

6

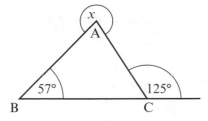

> Angles on a straight line add up to 180°, angles around a point add up to 360° and angles in a triangle add up to 180°.

Work out the size of the angle marked *x*

Angle ACB = ... − 125° = ...°

Angle BAC = 180° − 57° − ....° = ...°

x = 360° − ....° = ...°   **(3 marks)**

7 The diagram shows a five-sided shape. All the sides are equal in length.

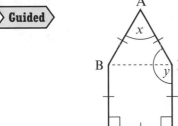

(a) Work out the size of the angle marked *x*
Give a reason for your answer.

.... ÷ ....

= ....°   **(1 mark)**

Triangle ABE is an .... triangle.   **(1 mark)**

(b) Work out the size of the angle marked *y*

...............................°   **(2 marks)**

8 Work out the size of each marked angle.
Give a reason for your answers.

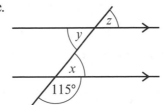

> Angles *x* and *y* are alternate angles between parallel lines.

(a) *x* = ...°

Reason .... **(2 marks)**

(b) *y* = ...°

Reason .... **(2 marks)**

(c) *z* = ...°

Reason .... **(2 marks)**

Solving angle problems

 1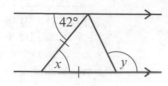

 Guided

(a) Work out the size of the angle marked x

$x =$ ....° **(1 mark)**

(b) Give a reason for your answer.

.... angles are equal **(1 mark)**

(c) Work out the size of the angle marked y

$180° - x =$ ....

.... $\div 2 =$ ....

$y = 180° -$ .... $=$ .... **(3 marks)**

2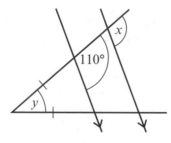

(a) Work out the size of the angle marked x

....................................° **(1 mark)**

(b) Give a reason for your answer.

.... **(1 mark)**

(c) Work out the size of the angle marked y

....................................° **(2 marks)**

(d) Give a reason for your answer.

.... **(1 mark)**

 3 The diagram shows 3 straight lines. Work out the value of x

 Guided

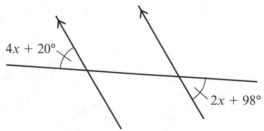

Alternate angles are equal.

Corresponding angles are equal.

$4x + 20 =$ ....

Solve for x.

....................................° **(3 marks)**

Angles in polygons

1 The diagrams show regular polygons.
Work out the size of an exterior angle for each regular polygon.

(a)

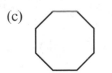

Exterior angle = 360 ÷ number of sides

Exterior angle = 360° ÷ .... = ...° **(2 marks)**

(b)

(c)

..........................° **(2 marks)**　　　　　　　　　　　　　..........................° **(2 marks)**

2 The interior angle of a regular polygon is 140°.

(a) Work out the size of the exterior angle of the polygon.

Exterior angle = 180 − .... = ...° **(1 mark)**

(b) Work out the number of sides of the polygon.

Number of sides − 360 ÷ .... = ... **(2 marks)**

3 Each diagram shows part of a regular polygon.
The size of one interior angle is given.
Work out the number of sides in each polygon.

You will need to use problem-solving skills throughout your exam – **be prepared!**

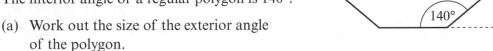

(a)

150°

(b)

144°

(c)

162°

.... **(3 marks)**　　.......................... **(3 marks)**　　.......................... **(3 marks)**

4 The diagram shows three sides of a regular hexagon. Show that $x = 30°$.

(3 marks)

5 The sum of the interior angles of a regular polygon is 2340°.
(a) Work out the number of sides of the polygon. **(3 marks)**

(b) Work out the size of an exterior angle of the polygon. **(2 marks)**

Time and timetables

Guided

1 Write these times using the 24-hour clock.

(a) 3.15 pm   (b) 2.25 am   (c) 11.48 pm

.....................:15   **(1 mark)**   ....:25   **(1 mark)**   ....   **(1 mark)**

2 These times are given using the 24-hour clock. Write the times using am or pm.

(a) 04:25   (b) 12:10   (c) 20:32

....   **(1 mark)**   ....   **(1 mark)**   ....   **(1 mark)**

3 A train sets off from Coventry at 08:25 and arrives at London Euston at 09:17. How long does the journey take?

Guided

08.25   to   09.00   to   09.17

> Break the journey up.

.... mins + ... mins = ... mins

.... minutes   **(2 marks)**

4 A cyclist sets off at 10.25 am and cycles for one and three quarter hours. He then rests for 35 minutes and returns home by a different route which takes 2 hours and 45 minutes. What time does he arrive back home?

> Give your answer using the 24-hour clock.

....   **(2 marks)**

5 Here is part of a timetable.

Train	A	B	C	D	E
Wolverhampton	06:45	07:05	07:25	07:45	08:10
London	08:35	08:56	09:15	09:34	10:15

(a) Which train took more than two hours to go from Wolverhampton to London?

....   **(1 mark)**

(b) Work out the number of minutes taken by train B to go from Wolverhampton to London.

.... minutes   **(2 marks)**

(c) Aaron lives in Wolverhampton and has a meeting in London. He needs to arrive in London before 10:00. Write down the time of the latest train he can catch.

....   **(1 mark)**

(d) Train B arrives 36 minutes late in London. What time does this train arrive in London?

....   **(1 mark)**

Reading scales

1 Write the number marked with an arrow.

Look at the scales carefully.

 Guided

(a)

2.................... **(1 mark)**

(b)

3.................... **(1 mark)**

(c)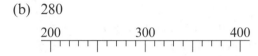

.... **(1 mark)**

(d)

.... **(1 mark)**

2 Mark each number with an arrow (↑) on the number line.

(a) 33

(1 mark)

(b) 280

(1 mark)

(c) 5700

(1 mark)

(d) 3.8

(1 mark)

3 (a) Write the speed marked with an arrow.

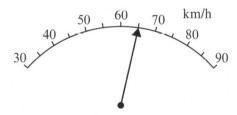

.........................km/h **(1 mark)**

(b) Mark 42 km/h with an arrow (↑).

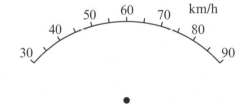

(1 mark)

4 The diagram shows six identical cubes and four identical triangles on two scales.

 PROBLEM SOLVED!

You will need to use problem-solving skills throughout your exam – **be prepared!**

Work out the mass, in kg, of one triangle.

.........................kg **(3 marks)**

Perimeter and area

1 A shape has been drawn on a grid of centimetre squares.

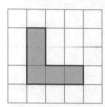

(a) Work out the area of the shape.

> Count the number of shaded squares to find the area.

....................cm²   **(1 mark)**

(b) Work out the perimeter of the shape.

> Count around the shape to find the perimeter.

....................cm   **(1 mark)**

2 Two shapes have been drawn on a grid of centimetre squares.

> Count 1 cm² for every whole square and $\frac{1}{2}$ cm² for every part square.

(a)

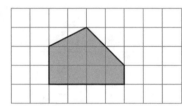

Work out the area of the shape.

Area of whole squares = ...cm²

Area of part squares = ...cm²

Total area = ... + ... = ...cm²   **(1 mark)**

(b)

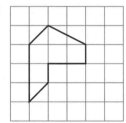

Work out the area of the shape.

....................cm²   **(1 mark)**

3 Shape X is a trapezium, two of its angles are right angles.
A rectangle, R, is cut from X to form shape Y.

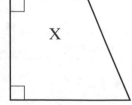 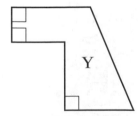

Tick a box to say which **one** of these statements is true.   **(1 mark)**

The perimeter of X is longer than the perimeter of Y. ☐

The perimieter of X is the same as the perimeter of Y. ☐

The perimeter of X is shorter than the perimeter of Y. ☐

It is **not** possible to tell which perimeter is longer. ☐

Area formulae

1   Work out the areas of the following shapes.

(a)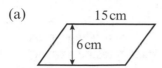

Area = ... × ....

Area = ...cm²   **(2 marks)**

(b)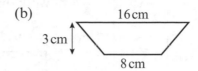

Area = ½(......... + ...) × ....

Area = ...cm²   **(2 marks)**

(c)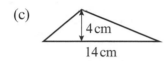

Area = ½ × .... × ....

Area = ...cm²   **(2 marks)**

(d)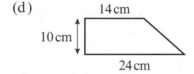

Area = ½(......... + ...) × ....

Area = ...cm²   **(2 marks)**

2   The trapezium has an area of 12.6 cm².

Work out the length x

$\frac{1}{2}$ × 2.8 × (................. + ...) = 12.6

1.4 × (................. + ...) = 12.6

.... + ... = 12.6 ÷ ....

x = ...

(3 marks)

3   The area of the parallelogram is three times the area of the trapezium.

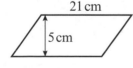   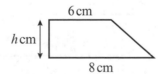

Work out the height, h cm, of the trapezium. You **must** show your working.

.... **(3 marks)**

Solving area problems

1

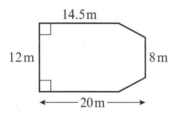

The diagram shows a compound shape.
Work out the area of the shape.

> **Guided**

| A compound shape is a shape that can be divided into two or more simple shapes. | Divide the shape into a rectangle and a triangle. Label the shapes A and B. |

Area A = ... × .... = ...cm²

Area B = $\frac{1}{2}$ × .... × .... = ...cm²

Total area = Area A + Area B = ... + ...

= ....cm²   **(3 marks)**

2

14.5 m

12 m   8 m

◄——— 20 m ———►

The diagram shows the plan of a car park.
The council wants to sell the car park.
The council wants at least £27 per square metre.
A local developer offers £6000. Will the council accept this offer?

.... **(5 marks)**

3 This shape has an area of 102 cm².

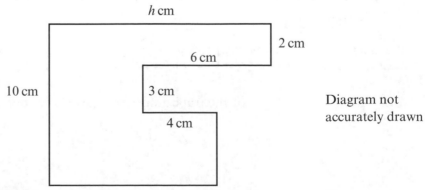

h cm

2 cm

6 cm

10 cm   3 cm

4 cm

Diagram not accurately drawn

Work out the length h
You **must** show your working.

(4 marks)

3D shapes

1 Write the names of these shapes.

(a)

Cube **(1 mark)**

(b)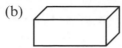

Cub.................... **(1 mark)**

(c)

.... **(1 mark)**

(d)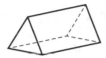

.... **(1 mark)**

(e)

.... **(1 mark)**

(f)

Sp...................... **(1 mark)**

2 A 3D shape has five faces, eight edges and five vertices.
Circle the name of this shape.

Cuboid   Triangular prism   Square-based pyramid   Tetrahedron   **(1 mark)**

3 Work out the total surface area of each shape.

(a)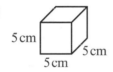

Area of 1 face = ... × ....cm²

Area of 6 faces = 6 × .... = ...cm²   **(2 marks)**

(b)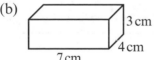

...........................cm² **(2 marks)**

(c)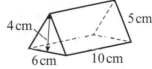

...........................cm² **(2 marks)**

4 The diagram shows a cube.

(a) Work out the total surface area of the cube.

......................................cm² **(2 marks)**

(b) Sara wants to paint all 6 faces of 40 of these cubes.
Each tin of paint covers an area of 350 cm². She buys ten tins.
Does she buy enough tins?
Give a reason for your answer.

.... **(3 marks)**

83

Volumes of cuboids

1 Work out the volumes of these cuboids.

(a)

5 cm
5 cm
5 cm

Volume = length × width × height

Volume = ... × .... × .... = ... cm³   **(2 marks)**

(b)

6 cm
15 cm
4 cm

.... cm³   **(2 marks)**

(c)

16 cm
12 cm
7 cm

.... cm³   **(2 marks)**

2 A cuboid has a volume of 504 cm³, a length of 12 cm and a width of 7 cm.
Work out the height of the cuboid.

Volume = length × width × height

.... = ... × .... × height

height = ... ÷ .... = ... cm   **(3 marks)**

3 A box measures 175 cm × 120 cm × 90 cm. The box is to be completely filled
with cuboids. Each cuboid measures 25 cm × 20 cm × 10 cm.
Work out the number of cuboids which can completely fill the box.

> Work out how many cuboids can fit along each dimension of the box.

= .... cuboids   **(3 marks)**

4 The diagrams show a rectangular tray and a carton.

> You will need to use
> problem-solving skills
> throughout your exam
> – **be prepared!**

70 cm
3 cm
50 cm
30 cm

The rectangular tray has length 70 cm, width 50 cm and depth 3 cm.
The carton is a cuboid with a square base of side 30 cm.
The tray is full of water. The water is poured into the empty carton.
Work out the depth, in cm, of the water in the carton.

.... cm   **(3 marks)**

Prisms

1 Work out the volumes of these prisms.

(a)

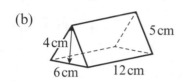

> Volume of a prism = area of cross-section × length.
> You need to learn this formula for your exam.

Volume = $\left(\frac{1}{2} \times \text{...............} \times \text{...............}\right) \times \text{...............} = \text{...............} cm^3$ **(3 marks)**

(b)

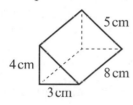

(c)

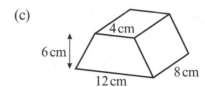

...........................cm³   **(3 marks)**   ....cm³ **(3 marks)**

2 Work out the total surface area of each shape.

> The total surface area is found by adding together the areas of all the faces.

(a)

Surface area = $2\left(\frac{1}{2} \times \text{............} \times \text{............}\right) + \left(\text{............} \times \text{............}\right)$

$+ \left(\text{............} \times \text{............}\right) + \left(\text{............} \times \text{............}\right)$

Surface area = ...cm²   **(3 marks)**

(b)

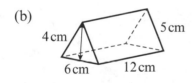

(c)

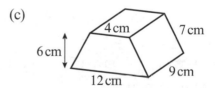

...........................cm²   **(3 marks)**   ....cm² **(3 marks)**

3 The diagram shows a prism.
Work out the volume of the prism.

> The shaded area is a compound shape.

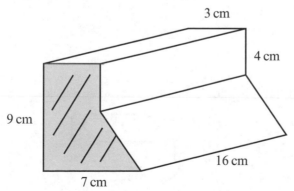

(3 marks)

Units of area and volume

1 Convert

(a) 6 m² into cm²

$\boxed{1\,m^2 = 100 \times 100\,cm^2}$

6 m² = 6 × .... × .... = ... cm²   **(1 mark)**

(b) 15 cm² into mm²   (c) 4 km² into m²   (d) 500 000 cm² into m²

.... mm² **(1 mark)**   .... m² **(1 mark)**   .... m²   **(1 mark)**

(e) 60 000 mm² into cm²   (f) 800 000 m² into km²   (g) 275 mm² into cm².

.... cm² **(1 mark)**   .... km² **(1 mark)**   .... cm² **(1 mark)**

2 Convert

(a) 22 m³ into cm³

$\boxed{1\,m^3 = 100 \times 100 \times 100\,cm^3}$

22 m³ = 22 × .... × .... × .... = ... cm³   **(2 marks)**

(b) 28 cm³ into mm³   (c) 3 km³ into m³   (d) 200 000 000 cm³ into m³

.... mm³ **(2 marks)**   .... m³ **(2 marks)**   .... m³   **(2 marks)**

(e) 50 000 000 mm³ into cm³   (f) 200 000 cm³ into litres   (g) 8 m³ into litres.

.... cm³ **(2 marks)**   .... litres **(2 marks)**   .... litres **(2 marks)**

3 Dan is trying to convert 25 m³ into litres.

This is his working:

25 m³ = 25 × 100 × 100 cm³ = 250 000 cm³ = 250 000 ÷ 1000 litres = 250 litres

Is Dan correct?

Show working to justify your answer.

(2 marks)

4 Work out how many litres of water each tank in the shape of a cuboid can hold.

(a)

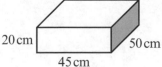

(b)

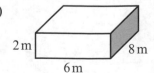

.... litres   **(2 marks)**   .... litres   **(2 marks)**

Translations

1   Write the following information in vector notation.

(a)   3 to the right
      4 up

$\begin{pmatrix} 3 \\ \rule{1cm}{0.4pt} \end{pmatrix}$ **(1 mark)**

(b)   2 to the left
      3 down

$\begin{pmatrix} - \\ \rule{1cm}{0.4pt} \end{pmatrix}$ **(1 mark)**

(c)   5 to the left
      6 up

$\begin{pmatrix} \rule{1cm}{0.4pt} \\ \rule{1cm}{0.4pt} \end{pmatrix}$ **(1 mark)**

2   (a)   Translate shape A by the vector $\begin{pmatrix} 3 \\ -4 \end{pmatrix}$
          Label the image C.

      (2 marks)

(b)   Translate shape B by the vector $\begin{pmatrix} -5 \\ 7 \end{pmatrix}$
      Label the image D.

      (2 marks)

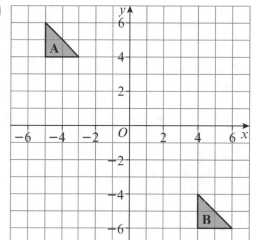

3   (a)   Describe fully the single transformation that will map shape G onto shape H.

....

....

....

(b)   Describe fully the single transformation that will map shape M onto shape N.

      (2 marks)

.... **(2 marks)**

4   Shape P is translated by the vector $\begin{pmatrix} -7 \\ 2 \end{pmatrix}$ onto shape Q.

Shape Q is translated by the vector $\begin{pmatrix} -1 \\ -5 \end{pmatrix}$ onto shape R.

Circle the vector that represents the single translation that maps shape P directly onto shape R.

$\begin{pmatrix} 7 \\ -10 \end{pmatrix}$ $\begin{pmatrix} -8 \\ 3 \end{pmatrix}$ $\begin{pmatrix} -6 \\ -3 \end{pmatrix}$ $\begin{pmatrix} -8 \\ -3 \end{pmatrix}$ **(1 mark)**

Had a go ☐ **Nearly there** ☐ **Nailed it!** ☐

Reflections

1   Reflect the shape in the mirror line (indicated by a dashed line).

> The reflected shape should be the same distance from the mirror line as the original shape. For diagonal mirror lines, measure distances diagonally.

(a)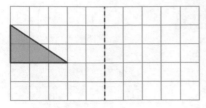

(1 mark)

(b)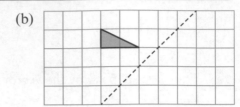

(1 mark)

2   (a) Reflect shape P in the line $y = 1$
        Label the image Q.

(2 marks)

> **Guided**

> First draw the line $y = 1$ on the graph.

(b) Reflect shape R in the line $x = 1$
     Label the image S.

(2 marks)

> First draw the line $x = 1$ on the graph.

(c) Reflect shape T in the line $y = x$
     Label the image U.

(2 marks)

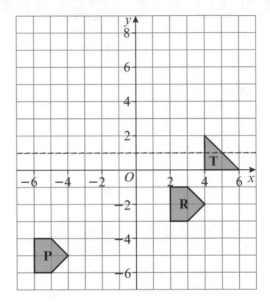

3   (a) Describe fully the single transformation that will map shape A onto shape B.

....

....

(2 marks)

(b) Describe fully the single transformation that will map shape B onto shape C.

....

....

(2 marks)

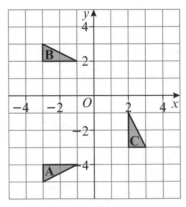

4   Reflect the triangle in the line $y = -x$

(2 marks)

Rotations

1 (a) Rotate triangle A 180° about the point (−1, 2).
Label the image B.

Guided

> Trace triangle A using tracing paper. Put your pencil at the point (−1, 2) and then rotate the tracing paper through 180°.

(2 marks)

(b) Rotate triangle A 90° anticlockwise about the point (2, 1).
Label the image C.

(2 marks)

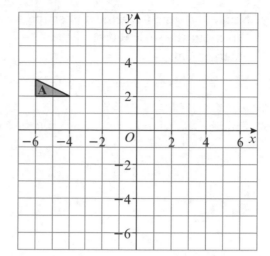

2 (a) Rotate shape P 90° clockwise about the point (0, 1).
Label the image Q.

(2 marks)

(b) Rotate shape P 180° about the point (−2, −1).
Label the image R.

(2 marks)

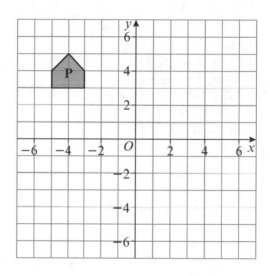

3 (a) Describe fully the single transformation that will map shape A onto shape B.

....

....

....

(b) Describe fully the single transformation that will map shape A onto shape C.

(3 marks)

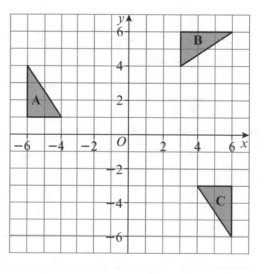

....

....

....

(3 marks)

Had a go ☐   Nearly there ☐   Nailed it! ☐

Enlargements

1 (a) Shape B is an enlargement of shape A.
Work out the scale factor of the
enlargement.

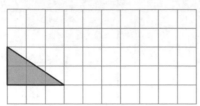

.... ÷ .... = ... **(1 mark)**

(b) Enlarge the triangle below by scale factor 2.

> No centre of enlargement is
> given so the enlarged shape can
> be placed anywhere on the grid.

(1 mark)

2 Enlarge shape A by scale factor 3,
centre (0, 0).
Label the image B.

> Work out the column vectors
> from the centre of enlargement
> to each point, and apply the
> scale factor.

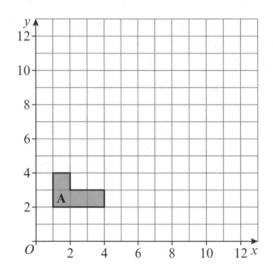

(2 marks)

3 (a) Describe fully the single
transformation that will map
shape A onto shape B.

> The image is **smaller** than the object,
> so the scale factor will be a **fraction**.

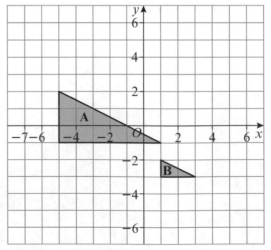

....

....

....

(3 marks)

(b) Enlarge shape A by scale factor $\frac{1}{2}$, centre (−7, −5).

(2 marks)

Pythagoras' theorem

Guided

1 Work out the lengths of the sides marked with letters in these triangles.

Give your answers to 3 significant figures.

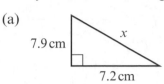

short^2 + short^2 = long^2

(a)

7.9 cm

x

7.2 cm

$x^2 = \ldots\ldots\ldots^2 + \ldots\ldots\ldots^2 = \ldots\ldots\ldots$

$x = \sqrt{\ldots\ldots\ldots\ldots\ldots\ldots}$

$x = \ldots\ldots\ldots\ldots$ cm

(2 marks)

(b)

5.6 cm

16.7 cm

z

$\ldots\ldots\ldots^2 = z^2 + \ldots\ldots\ldots^2$

$z^2 = \ldots\ldots\ldots^2 - \ldots\ldots\ldots^2 = \ldots\ldots\ldots$

$z = \sqrt{\ldots\ldots\ldots\ldots\ldots\ldots}$

$z = \ldots\ldots\ldots\ldots$ cm

(2 marks)

2 Cindy has a rectangular suitcase of length 95 cm and width 72 cm.

She wants to put her walking stick into her suitcase.

The length of the walking stick is 125 cm.

She thinks that the walking stick will fit into her suitcase. Is she correct?

Show working to justify your answer.

$\ldots\ldots\ldots\ldots\ldots\ldots\ldots\ldots\ldots\ldots\ldots\ldots\ldots\ldots\ldots\ldots\ldots\ldots\ldots\ldots\ldots\ldots\ldots\ldots\ldots$ **(2 marks)**

3 Here is a triangle.

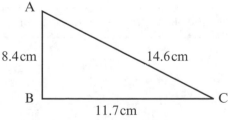

A

8.4 cm

B

11.7 cm

14.6 cm

C

Is the triangle right-angled?

Show working to justify your answer.

(3 marks)

Line segments

1 Find the length of the following line segment. Give your answer to 3 significant figures.

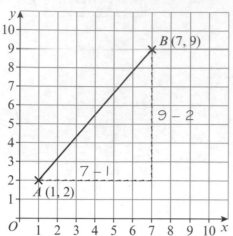

$AB^2 = $ ....² $+$ ....²

$AB^2 = $ ....

$AB = \sqrt{\text{.................}}$

$AB = $ ....

> Draw a horizontal and a vertical line to make a triangle.

(2 marks)

2 Find the length of the following line segment. Give your answer to 3 significant figures.

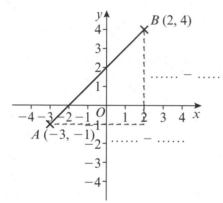

$AB^2 = $ ....² $+$ ....²

$AB^2 = $ ....

$AB = \sqrt{\text{.................}}$

$AB = $ ....

> There is no grid so use subtraction to work out the distance across and the distance down.

(2 marks)

3 Point A has coordinates (3, 1) and point B has coordinates (11, 7).
Work out the length of the line segment AB

> Draw a sketch showing both points in roughly the right positions.

.... **(2 marks)**

4 The points A(−2, −6) and B(4, 2) are the opposite ends of a diameter of a circle.

(a) Find the coordinates of the centre of the circle.

.... **(2 marks)**

(b) Show that the radius of the circle is 5 units.

(2 marks)

Trigonometry 1

Guided

1   Work out the size of each of the angles marked with letters.
Give each answer correct to 1 decimal place.

SOH CAH TOA

(a)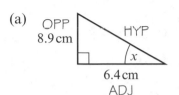

$\tan x = \dfrac{\text{opp}}{\text{adj}} = $ ....

....

$x = \tan^{-1}$ ....

$x = $ ....$°$ **(2 marks)**

> Start by labelling the sides of the triangle. Then write down the trigonometric ratio that uses the two given sides.

(b)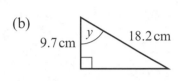

.... $y = \dfrac{\text{adj}}{\text{hyp}} = $ ....

....

$y = $ ....$^{-1}$ ....

$y = $ ....$°$ **(2 marks)**

2   Chris is trying to calculate angle x in this right-angled triangle.

This is his working.

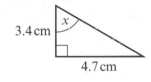

$\tan x = \frac{3.4}{4.7}$ $x = 35.882...$ $x = 35.9°$

(a) Without using your calculator how can you tell that Chris has made a mistake?

                 (1 mark)

(b) Work out the correct value of angle x **(1 mark)**

3   The diagram shows two right-angled triangles.
Work out the size of angle x.
Give your answer correct to 1 decimal place.

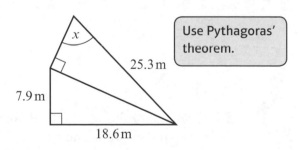

> Use Pythagoras' theorem.

....................................$°$ **(3 marks)**

4   The diagram shows a pitched roof.
Hayley wants to use smooth tiles to cover the roof.
The smooth tiles can only be used when the angle,
x, is no more than $17°$.
Can she use the smooth tiles on her roof?
Show working to justify your answer.

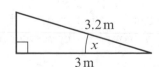

                 (3 marks)

Trigonometry 2

5 Work out the length, in cm, of each of the marked sides.
Give each answer to 1 decimal place.

> Guided

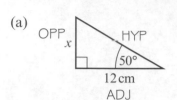

SOH CAH TOA

(a)

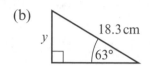

OPP x HYP
50°
12 cm
ADJ

$\tan \ldots\ldots° = \dfrac{\text{opp}}{\text{adj}} = \dfrac{x}{\ldots\ldots}$

$x = \ldots\ldots\ldots × \tan 50°$

$x = \ldots\ldots\ldots$ cm

(2 marks)

> Start by labelling the sides of the triangle. Then write the trigonometric ratio that uses the given and unknown side.

(b)

18.3 cm
y
63°

$\ldots\ldots 63° = \dfrac{\text{opp}}{\text{hyp}} = \dfrac{y}{18.3}$

$y = \ldots\ldots × \ldots\ldots 63°$

$y = \ldots\ldots$ cm

(2 marks)

6 A 6 m long ladder rests against a vertical wall.
The foot of the ladder rests on horizontal ground.
The ladder makes an angle of 73° with the ground
when it is leaning against the wall.
How far does the ladder reach up the wall?

> Guided

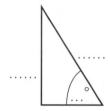

......
......
...°

$\ldots\ldots\ldots\ldots\ldots\ldots\ldots\ldots$ m   **(2 marks)**

7 A ladder leans against a vertical wall. The foot of the ladder is 1.9 m from the foot
of the wall and makes an angle of 75° with the horizontal ground.

How long is the ladder?

> You are calculating the hypotenuse so be careful with your calculation.

(2 marks)

8 The diagram shows a vertical pole standing on
horizontal ground. The points A, B and C are
in a straight line on the ground. The point D
is at the top of the pole so that DC is vertical.
The angle of elevation of D from A is 35°.

(a) Work out the height of the pole.
Give your answer correct to 1 decimal place.

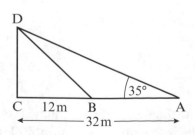

D
C   12 m   B   35°   A
←——— 32 m ———→

$\ldots\ldots\ldots\ldots\ldots\ldots\ldots\ldots$ m   **(2 marks)**

(b) Work out the size of the angle of elevation of D from B.
Give your answer correct to 1 decimal place.

$\ldots\ldots\ldots\ldots\ldots\ldots\ldots\ldots$°   **(2 marks)**

Solving trigonometry problems

These are non-calculator problems.

1 Complete the table.

You must remember these for the exam.

	0°	30°	45°	60°	90°
sin	....	$\frac{1}{2}$	....	....	....
cos	....	....	$\frac{1}{\sqrt{2}}$	....	....
tan	....	....	....	$\sqrt{3}$	⟶ ∞

(5 marks)

2 Work out the length, in cm, of side x

SOH CAH TOA

Guided

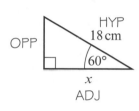

$\cos 60° = \dfrac{adj}{hyp} = \dfrac{...................}{...................}$

$x = ... \times \cos 60°$

$x = ...\,cm$

(2 marks)

Start by labelling the sides of the triangle. Then write down the trigonometry ratio that uses the given and unknown side.

3 Work out the size of each of the angles marked with letters.

Guided

(a)

$\tan x = \dfrac{.............}{.............} = \dfrac{...........}{...........}$

$x = ...°$

(2 marks)

You can use the answers to question 1 here.

(b)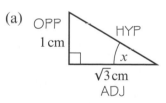

$............... y = \dfrac{...................}{...................} = \dfrac{...................}{...................}$

$y = ...°$

(2 marks)

4 Alan is sitting on the ground and his distance from the base of a tower is 30 feet. The angle of elevation from Alan to the top of the tower is 60°. Work out the height of the tower.
Give your answer as an exact value.

Draw a diagram.

....................................feet **(2 marks)**

5 A flagpole is 22 m high and is supported by a cable making an angle of 30° to the flagpole.

Work out the length of the cable, marked x on the diagram.
Give your answer as an exact value.

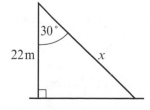

(2 marks)

Measuring and drawing angles

1 Measure the sizes of these angles.

First estimate the size of the angle.

(a)

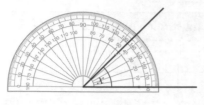

..................................° **(1 mark)**

(b)

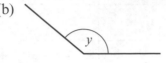   *y*

..................................° **(1 mark)**

(c)

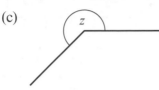   *z*

Measure the smaller angle and then subtract it from 360°.

..................................° **(1 mark)**

2 Accurately draw these angles.

(a) 74°   (b) 148°

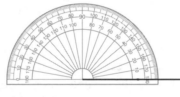

(1 mark)   **(1 mark)**

3 Katie wants to draw a reflex angle of 294°.
She has a protractor with a scale going from 0° to 180°.
What angle should she actually draw using her protractor?
Circle your answer.

24°   66°   106°   104°   **(1 mark)**

4 Measure and name these angles.

To name the angles, choose from acute, obtuse and reflex.

(a)   (b)   (c)

 x   *y*   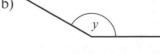 *z*

.... **(1 mark)**   .... **(1 mark)**   .... **(1 mark)**

5 Make an accurate drawing of triangle ABC, where BC = 5 cm, angle ABC = 58° and angle ACB = 44°.

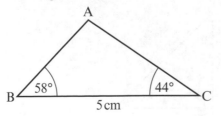

(2 marks)

Measuring lines

1 Measure the lengths of these lines.
State the units of measurement.

First estimate the length of the line.

(a)

(b)

.... **(1 mark)**   .... **(1 mark)**

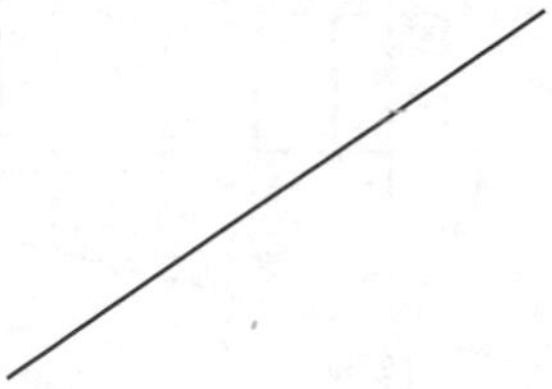

(c) ——————————————

.... **(1 mark)**

2 In the space, draw straight lines with these lengths.

(a) 52 mm   (b) 6 cm   (c) 7.8 cm

(1 mark)   **(1 mark)**   **(1 mark)**

3 Mark the midpoint of the line AB with a cross (×).

A ————————————————————————————— *B*   **(2 marks)**

4 The diagram shows an adult woman standing next to a building.
The woman and the building are drawn to the same scale.
Work out an estimate for the height, in metres, of the building.

1.6 m is a good estimate for the height of an adult woman.

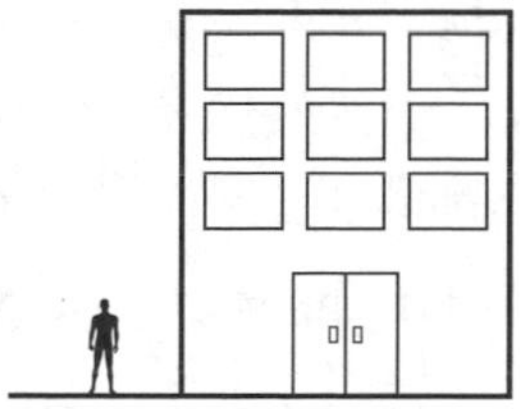

.... m   **(2 marks)**

5 The picture shows a house.

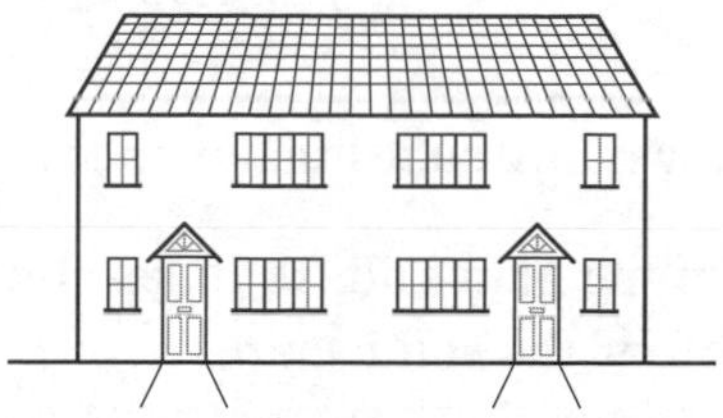

(a) Write an estimate for the height, in metres, of the front door.

.... m

(b) Use your estimate from part (a) to write an estimate for height, in metres, of the house.

(1 mark)

.... m   **(1 mark)**

Plans and elevations

1   Match each solid shape to its net.

Guided

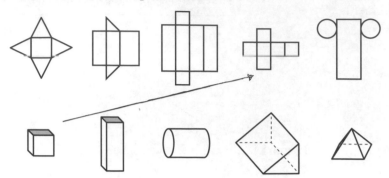

When the net (arrowed) is folded it will be a cube.

(3 marks)

2   The diagram shows a solid object made of eight identical cubes.

Guided

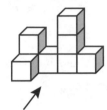

On the grids, draw the front elevation and the plan of the solid object.

Front elevation   Plan   **(2 marks)**

3   Here are the plan, front elevation and side elevation of a 3D shape.

plan   front elevation   side elevation

(a)   On the isometric paper, draw a sketch of the 3D shape.

(2 marks)

(b)   How many cubes make up this 3D shape?   .... **(1 mark)**

4   Here are the plan and front elevation of a solid shape. On the third grid, draw the side elevation of the solid shape.

Plan   Front elevation   **(2 marks)**

GEOMETRY & MEASURES

Scale drawings and maps

1 The lines are drawn to scale.
Work out the actual lengths by using the scales.

 Guided

Measure the line first.

(a) 1 cm to 10 m

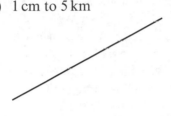

Actual length = ... × 10 m = ... m   **(2 marks)**

(b) 1 cm to 5 km

(c) 1 cm to 15 km

.........................km   **(2 marks)**   ....km   **(2 marks)**

 Guided

2 On a map the distance between two towns is measured and recorded.
Work out the actual distance between the towns using the scale shown.

(a) Distance on map = 5 cm
Scale = 1 : 50 000

Actual distance = ... × 50 000 = ... cm

= .... ÷ 100 = ... m

= .... ÷ 1000 = ... km   **(2 marks)**

(b) Distance on map = 12 cm
Scale = 1 : 100 000

(c) Distance on map = 15.4 cm
Scale = 1 : 1 000 000

.........................km   **(2 marks)**   ....km   **(2 marks)**

3 What distance on a map will represent an actual distance of

Convert 10 km into cm.

(a) 10 km using a scale
1 : 50 000

(b) 15 km using a scale
1 : 100 000

(c) 50 km using a scale
1 : 1 000 000?

.................cm   **(2 marks)**   ....cm   **(2 marks)**   ....cm   **(2 marks)**

4 Arthur uses a scale of 1 : 300 to make a model of an aeroplane.
The length of the real aeroplane is 69 m.
Which of these is the length of the model?

Circle your answer.

2.3 mm   23 mm   230 mm   2300 mm   **(1 mark)**

Constructions 1

1 Use a ruler and compasses to construct the perpendicular bisector of *AB*

> **Guided**

> Do not rub out the arcs you make when using your compasses.

A ———————————— *B*

> 1. Draw an arc, centre *A*, with radius more than half the length of *AB* above and below the line segment *AB*.
> 2. Draw another arc, centre *B*, with the same radius above and below the line segment *AB*.
> 3. Draw a line through the two points where the arcs cross each other above and below the line segment *AB*.

(2 marks)

2 Use a ruler and compasses to construct the perpendicular to the line segment AB that passes through the point T.

> **Guided**

> You must show all your construction lines.

> Draw two arcs, centre *T*, with the same radius to cross either side of *T*.

A ——×——————— *B*
　　　T

(2 marks)

3 Construct the line that is the shortest distance between *P* and the line *AB*

× *P*

A ———————— *B*

(3 marks)

Constructions 2

4 Use a ruler and compasses to construct a triangle with sides of lengths 3.5 cm, 4 cm and 5 cm.

Guided

A ————————— B

> Do not rub out the arcs you draw when using your compasses.

(2 marks)

> 1. Draw a horizontal line of 5 cm and label it *AB*.
> 2. Set the compasses at 3.5 cm then draw an arc with centre *A*.
> 3. Set the compasses at 4 cm then draw an arc with centre *B*.
> 4. Draw lines from the point of intersection to *A* and *B*.

5 Use a ruler and compasses to construct the bisector of angle ABC

Guided

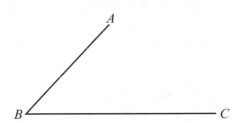

> 1. Draw an arc, centre *B*, to cross *AB* at *P* and *BC* at *Q*.
> 2. Draw an arc, centre *P*, and an arc, centre *Q*, with the same radius. The two arcs intersect.
> 3. Draw a line through point *B* and the point of intersection.

(2 marks)

6 Use a ruler and compasses to construct a 60° angle at *A*
You must show all your construction lines.

A ————————— B

(2 marks)

7 Use a ruler and compasses to construct a 45° angle at *A*
You must show all your construction lines.

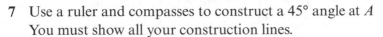

A ————————— B

(3 marks)

Loci

1 Draw the locus of all points that are exactly 2 cm from the line *AB*

A ——————————— B

> 1. Draw a circle of radius 2 cm with centre *A*.
> 2. Draw a circle of radius 2 cm with centre *B*.
> 3. Draw two parallel lines 2 cm above and below the line *AB*.

(2 marks)

2 The diagram shows the boundary of a rectangular garden, *ABCD*
A dog is tied to corner *B* with a rope of length 6 m.
Shade the region where the dog can reach.

1 cm represents 2 m.

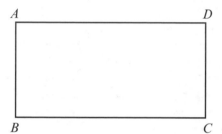

> 1. Use the scale to set the compasses at the required distance.
> 2. Draw an arc with centre *B*.
> 3. Shade the required region.

(2 marks)

3 *P*, *Q* and *R* represent three radio masts on a plan. Signals from mast *P* can be received 125 km away, from mast *Q* 75 km away and from mast *R* 50 km away. Show, by shading, the region in which signals can be received from all three masts.

1 cm represents 25 km.

× *Q*

P ×

× *R*

(3 marks)

4 *ABC* is a triangle. Shade the region inside the triangle which is both less than 3 cm from point *B* and closer to line *AC* than *AB*

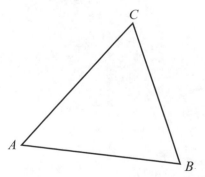

(4 marks)

Bearings

1 What is the angle between these starting and finishing positions?

(a) Face north-east, travel clockwise to face north-west. .... **(1 mark)**

(b) Face south, travel clockwise to face north-east. .... **(1 mark)**

(c) Face west, travel anticlockwise to face south-east. .... **(1 mark)**

2 Work out the bearing of

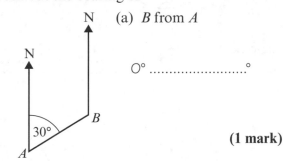

(a) *B* from *A*

O° ....°

(1 mark)

(b) *A* from *B*

180° + ...°

= ....°

(1 mark)

Guided

3 Work out the bearing of

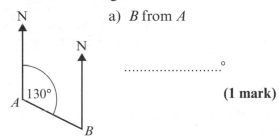

a) *B* from *A*

........................°

(1 mark)

(b) *A* from *B*

........................°

(1 mark)

4 Draw a line on a bearing of

(a) 030°   (b) 260°   (c) 320°.

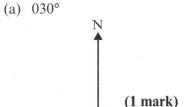 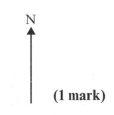 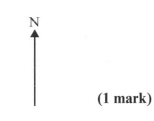

(1 mark)   **(1 mark)**   **(1 mark)**

5 The diagram shows three locations on a map. The scale of the map is 1 cm to 4 km.

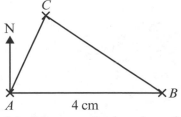

(a) Measure the bearing of *C* from *A* ....° **(1 mark)**

D is a fourth location. The actual distance of *D* from *B* is 20 km.
The bearing of *D* from *B* is 075°.

(b) Mark with a cross (✕) the position of *D* on the diagram. Label the point *D* **(2 marks)**

6 The bearing of *Q* from *P* is 245°.
Which of these is the bearing of *P* from *Q*?
Circle your answer.

> Draw a diagram.

025°   065°   115°   245° **(1 mark)**

103

Circles

1 Label the diagram with the correct names for the parts of a circle:

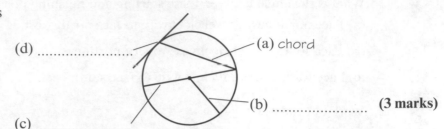

(d) ....

(a) *chord*

(b) .... **(3 marks)**

(c) ....

2 Work out the circumferences of these circles. Give your answers to 3 significant figures.

> You need to learn this formula.

(a)
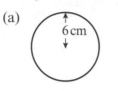
6 cm

$C = 2 \times \pi \times r$

$= 2 \times \pi \times$ ....

$=$ .... cm **(2 marks)**

(b)
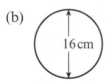
16 cm

> Check whether you are given the **radius** or the **diameter**.

.... cm **(2 marks)**

3 Work out the radii of circles with these circumferences. Give your answers to 3 significant figures.

(a) circumference = 35 cm   (b) circumference = 92 cm

$C = 2 \times \pi \times r$

.... $= 2 \times \pi \times r$

$r =$ .... ÷ ....

radius = ... cm **(2 marks)**   radius = ... cm **(2 marks)**

4 Work out the perimeters of these shapes. Give your answers to 3 significant figures.

(a)

12 cm

(b)
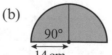
90°
14 cm

.... cm **(3 marks)**   .... cm **(3 marks)**

5 A reel of thread has a radius of 2.5 cm. The thread is wrapped round the reel 200 times. Work out the length of the thread. Give your answer to 3 significant figures.

.... cm **(3 marks)**

Area of a circle

1 Work out the areas of these circles.
Give your answers to 3 significant figures.

Guided

(a)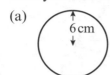

$A = \pi \times r^2$

You need to learn this formula.

$= \pi \times ....^2$

$= .... \ cm^2$ **(2 marks)**

(b)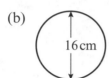

First work out the radius.

(c)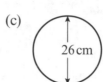

.....................................cm² **(2 marks)**   ....cm² **(2 marks)**

2 Work out the areas of these shapes.
Give your answers to 3 significant figures.

Guided

(a)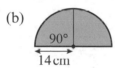

Area of whole circle $= \pi \times r^2$

$= \pi \times ....^2$

$= .... \ cm^2$

Area $= .... \div 4 = ... \ cm^2$ **(3 marks)**

(b)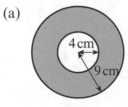

(c)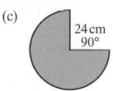

.....................................cm² **(3 marks)**   ....cm² **(3 marks)**

3 Work out the shaded area of each shape.
Give your answers to 3 significant figures.

(a)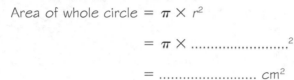

(b)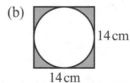

...................cm² **(3 marks)**   .... cm² **(3 marks)**

4 The diagram shows two identical squares of side 10 cm.

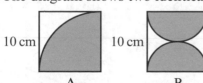

Diagram A shows a quarter of a circle shaded inside the square.
Diagram B shows two identical semi-circles shaded inside the square.

Show that the area of the region shaded in diagram A is equal to the area of the region shaded in diagram B.

(3 marks)

Sectors of circles

1 Work out the arc lengths of these sectors of circles. Give your answers to 3 significant figures.

> The arc length is the curved length.

Guided

(a)

$$\text{Arc length} = \frac{\text{.............}}{360} \times 2 \times \pi \times r$$

$$= \frac{\text{.............}}{360} \times \text{.......................}$$

.................cm   **(3 marks)**

(b)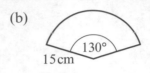

....cm   **(3 marks)**

2 Work out the perimeters of these sectors of circles. Give your answers to 3 significant figures.

> Find the arc length then add the two radii.

(a)

......................cm   **(4 marks)**

(b)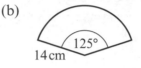

......................cm   **(4 marks)**

3 Work out the areas of these sectors of circles. Give your answers to 3 significant figures.

Guided

(a)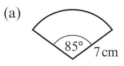

$$\text{Area of sector} = \frac{\text{.............}}{360} \times \pi \times r^2$$

$$= \frac{\text{.............}}{360} \times \pi \times \text{.....................} \times \text{.....................}$$

$$= \text{.........................} \text{ cm}^2$$

(3 marks)

(b)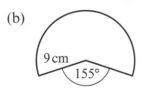

....cm²   **(3 marks)**

4 Here is a diagram of a sector of a circle of radius 8 cm.

Which of these is an expression for the arc length?

Circle your answer.

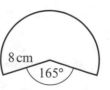

$\dfrac{195}{360} \times 2 \times \pi \times 8$ $\dfrac{165}{360} \times 2 \times \pi \times 8$ $\dfrac{195}{360} \times \pi \times 8 \times 8$ $\dfrac{165}{360} \times \pi \times 8 \times 8$ **(1 mark)**

5 The diagram shows a sector of radius 12 cm.
The area of the sector is 135 cm².
Work out the size of the angle marked x

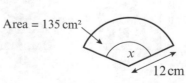

(3 marks)

Cylinders

1 Work out the volumes of these cylinders.
   Give your answers to 3 significant figures.

(a) 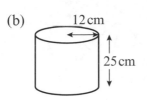 $V = \pi \times r^2 \times h$

4 cm

14 cm

$= \pi \times \text{............}^2 \times \text{............}$

$= \text{........................} \ cm^3$

(3 marks)

(b)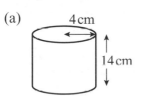

12 cm

25 cm

(c)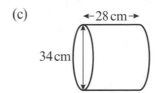

←28 cm→

34 cm

....cm³ **(3 marks)**   ....cm³ **(3 marks)**

2 Work out the total surface areas of these cylinders.
   Give your answers to 3 significant figures.

(a) 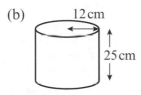   *Surface Area = (2 × π × r²) + (2 × π × r × h)*

4 cm

14 cm

$= (2 \times \pi \times \text{............}^2) + (2 \times \pi \times \text{............} \times \text{............})$

$= \text{........................} \ cm^2$

(3 marks)

(b)

12 cm

25 cm

(c)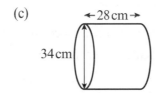

←28 cm→

34 cm

....cm² **(3 marks)**   ....cm² **(3 marks)**

3 Diagram A shows a cylinder with radius 15 cm
   and height 18 cm.
   Diagram B shows a cube with side 24 cm.

   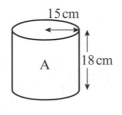 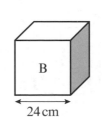

   15 cm

   A   18 cm

   B

   24 cm

   (a) Show that the volume of the cube is greater
       than the volume of the cylinder.

(4 marks)

   (b) Which of the shapes has the greater surface area?
       You **must** show your working.

(4 marks)

Volumes of 3D shapes

1 Work out the volumes of these shapes.
Give your answers to 3 significant figures.

(a)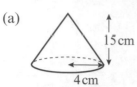
15 cm
4 cm

$V = \frac{1}{3} \times \pi \times r^2 \times h$

$= \frac{1}{3} \times \pi \times \text{............}^2 \times \text{............}$

$= \text{........................} \; cm^3$

(2 marks)

(b)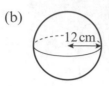
12 cm

....cm³ **(2 marks)**

(c)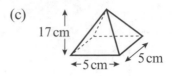
17 cm
5 cm   5 cm

....cm³ **(2 marks)**

2 A hemisphere has a diameter of 6 cm.
Ben works out the volume of the hemisphere.
This is his working.

6 cm

$\text{Volume} = \frac{4}{3} \times \pi \times r^3 = \frac{4}{3} \times \pi \times 6 \times 6 \times 6 = 288 \pi \, cm^3$

(a) Write down **two** mistakes that he has made.

Mistake 1 .... **(1 mark)**

Mistake 2 .... **(1 mark)**

(b) Work out the correct volume of the hemisphere.
Give your answer in terms of π.

288π is an example of an answer in terms of π.

(1 mark)

3 A cylinder has radius 4 cm and height 15 cm.
A cone has base radius 6 cm and height 10 cm.
Show that the volume of the cylinder is twice the
volume of the cone.

You will need to use problem-solving skills throughout your exam – **be prepared!**

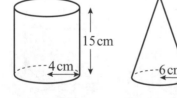
15 cm
4 cm
10 cm
6 cm

(4 marks)

4 Work out the volume of this solid.

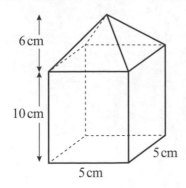
6 cm
10 cm
5 cm
5 cm

(3 marks)

Surface area

1 Work out the total surface areas of these shapes.
   Give your answers to 3 significant figures.

(a)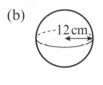

Total Surface Area = $(\pi \times r^2) + (\pi \times r \times l)$

= $(\pi \times \text{..............}^2)$

+ $(\pi \times \text{...............} \times \text{...............})$

= .... cm² **(2 marks)**

(b)

....cm² **(2 marks)**

(c)

....cm² **(2 marks)**

2 This solid is made from a cylinder of base
   radius 3 cm and height 5 cm, and a cone
   of slant height 21 cm.

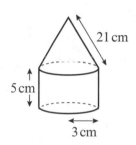

Don't forget the base of the cylinder.

Work out the total surface area of the solid.
Give your answer to 3 significant figures.   .... **(4 marks)**

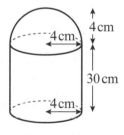

3 The diagram shows a solid made from a
   cylinder of base radius 4 cm and height 30 cm,
   and a hemisphere of radius 4 cm.

Work out the total surface area of the solid.
Give your answer to 3 significant figures.   .... **(4 marks)**

4 This cone has a base diameter of 12 cm and a
   perpendicular height of 8 cm.

PROBLEM SOLVED!

Work out the **total** surface area of the cone.
Give your answer in terms of π.

You will need to use problem-solving skills throughout your exam
– **be prepared!**

Use Pythagoras' theorem to work out the slant height.
Total surface area = curved surface area + area of base.

.... **(4 marks)**

Similarity and congruence

1   Here are five shapes.

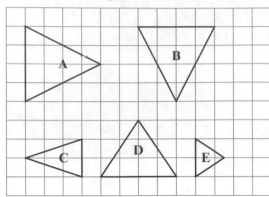

Congruent shapes have exactly the same size and shape.

(a)   Write the letters of two congruent shapes.

.... and ....   **(1 mark)**

Similar shapes are enlargements of one another.

(b)   Write the letters of two similar shapes.

.... and ....   **(1 mark)**

2   (a)   On the grid, draw a shape that is congruent to shape A.

Guided

(1 mark)

(b)   On the grid, draw a shape that is similar to shape B.

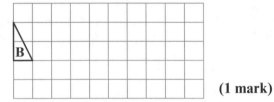

(1 mark)

3   On each grid, draw a shape that is similar to the shaded shape, but not congruent.

(a)

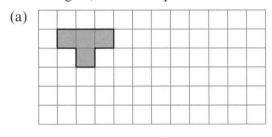

(1 mark)

(b)

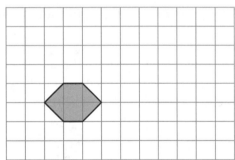

(1 mark)

4   These two cubes are similar. How many times will the 4 cm cube fit inside the 12 cm cube?

PROBLEM SOLVED!

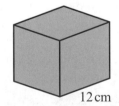

4 cm   12 cm

You will need to use problem-solving skills throughout your exam – **be prepared!**

....

(2 marks)

5   Which of the shapes A, B, C or D is congruent to shape X?

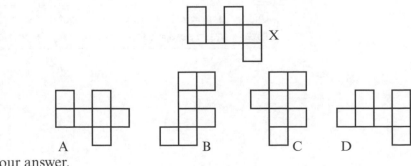

Circle your answer.

A   B   C   D   **(1 mark)**

Similar shapes

1 *ABC* and *PQR* are similar triangles.

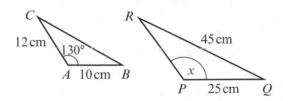

(a) What is the size of the angle marked *x*? ....° **(1 mark)**

(b) Work out the length of *PR*

$$\frac{PR}{12} = \frac{25}{\text{.............}}$$

$$PR = \frac{25}{\text{.............}} \times \text{.............}$$

$$= \text{.............................} \; cm$$ **(2 marks)**

> Use the fact that corresponding sides are in the same ratio.

> $10:25 = 2:5 = 12:....$

(c) Work out the length of *BC*

.... **(2 marks)**

2 The diagram shows two similar quadrilaterals.

(a) Work out the length of *AB*

$$\frac{AB}{\text{.............}} = \frac{\text{.............}}{\text{.............}}$$

$$AB = \frac{\text{.............}}{\text{.............}} \times \text{.............} = \text{.............} \; cm$$ **(2 marks)**

(b) Work out the length of *PS*

.... **(2 marks)**

3 The diagram shows two similar pentagons.

(a) Work out the value of *x*

x = ... **(2 marks)**

(b) Work out the value of *y*

y = ... **(2 marks)**

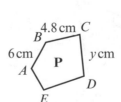

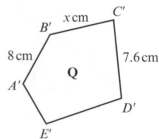

Congruent triangles

1 Show that triangle *ABC* is congruent to triangle *DEF*

> Guided

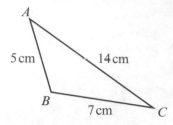

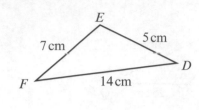

AC = DF

AB = ...

BC = ...

So the reason is ....

> You can write your reasons as SSS, ASA, SAS or RHS.

(3 marks)

2 Show that triangle *ABC* is congruent to triangle *PQR*

> Guided

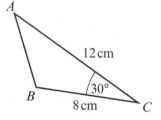

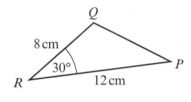

BC = ...

CA = ...

angle BCA = angle ....

So the reason is ....

(3 marks)

3 Which **two** of these triangles are congruent?
Circle the letters.

> Work out the missing angles.

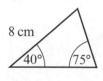 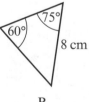 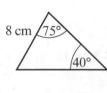   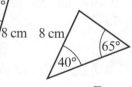

A   B   C   D   E   **(1 mark)**

Vectors

1 Write each vector as a column vector.

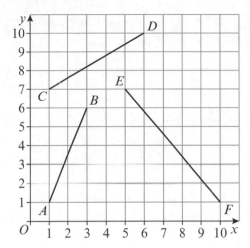

(a) $\overrightarrow{AB} = \begin{pmatrix} 2 \\ \cdots \end{pmatrix}$

(1 mark)

(b) $\overrightarrow{BA} = \begin{pmatrix} \cdots \\ -5 \end{pmatrix}$

(1 mark)

(c) $\overrightarrow{CD} = \begin{pmatrix} \cdots \\ \cdots \end{pmatrix}$

(1 mark)

(d) $\overrightarrow{DC} = \begin{pmatrix} \cdots \\ \cdots \end{pmatrix}$

(1 mark)

(e) $\overrightarrow{EF} = \begin{pmatrix} \cdots \\ \cdots \end{pmatrix}$

(1 mark)

(f) $\overrightarrow{FE} = \begin{pmatrix} \cdots \\ \cdots \end{pmatrix}$

(1 mark)

2 Vectors **a** and **b** are defined as

$a = \begin{pmatrix} -4 \\ 6 \end{pmatrix}$ $b = \begin{pmatrix} 7 \\ -2 \end{pmatrix}$

Write these vectors as column vectors.

(a) $5a$ **(1 mark)**

(b) $a + 2b$ **(3 marks)**

(c) $3a - b$ **(3 marks)**

3 $ABCD$ is a parallelogram.
AB is parallel to DC
AD is parallel to BC
Write these vectors
in terms of **p** and **q**

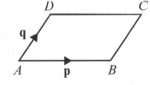

$\overrightarrow{DC} = \overrightarrow{AB} = p$ and $\overrightarrow{BC} = \overrightarrow{AD} = q$
So $\overrightarrow{AC} = \overrightarrow{AD} + \overrightarrow{DC} = \ldots\ldots\ldots\ldots$

(a) $\overrightarrow{AC}$ (b) $\overrightarrow{CA}$ (c) $\overrightarrow{DB}$ (d) $\overrightarrow{BD}$

....

(1 mark)   **(1 mark)**   **(1 mark)**   **(1 mark)**

4 Here is a vector triangle.

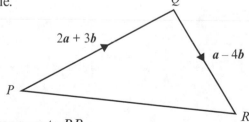

Circle the vector that represents RP

     $7b - a$ $a + 5b$ $b - 3a$ $3a - b$ **(1 mark)**

Problem-solving practice 1

1 *BEG* and *CFG* are straight lines. *ABC* is parallel to *DEF*. Angle *BEF* = 48°.
Angle *BCF* = 30°.

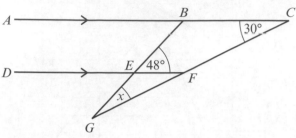

Work out the size of the angle marked *x*
Give reasons for each step of your working.

....° **(4 marks)**

2 A rectangular tray has length 60 cm, width 40 cm and depth 2 cm.
It is full of water. The water is poured into an empty cylinder of diameter
18 cm and height 20 cm. Will there be any water left in the rectangular tray?
You **must** show your working.

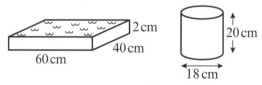

(4 marks)

3

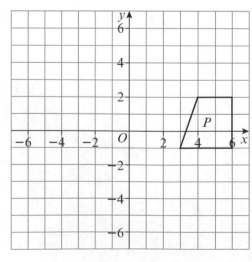

(a) Reflect shape *P* in the line *y* = −*x*
Label the image *Q*. **(2 marks)**

(b) Rotate shape *Q* 180° about the
point (0, 0). Label the image *R*. **(2 marks)**

(c) Describe fully the single transformation
that will map shape *P* onto shape *R*. **(2 marks)**

Problem-solving practice 2

4 The diagram shows a circular garden patio with a radius of 8 m. It has two small circles inside the garden patio. The small circles have a radius of 2 m. Kelly wants to gravel the shaded area. She orders 20 bags of gravel. Each bag covers an area of 9 m². Does she order enough bags to cover the shaded area? You **must** show your working.

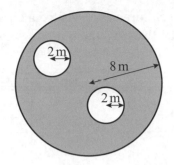

(5 marks)

5 This pattern is made from two **similar** trapeziums.

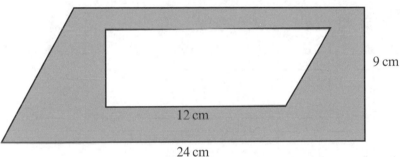

18 cm

9 cm

12 cm

24 cm

Show that the shaded area is 105 cm². **(4 marks)**

6 The diagram represents a vertical flagpole, AB. The flagpole is supported by two ropes, BC and BD, fixed to the horizontal ground at C and D.
$AB = 12.8$ m, $AC = 6.8$ m, angle $BDA = 42°$.

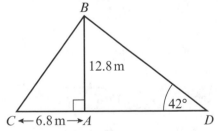

B

12.8 m

42°

C ← 6.8 m → A   D

(a) Work out the size of angle BCA
Give your answer to 3 significant figures.

....° **(2 marks)**

(b) Sandeep wants to replace rope BD. He buys a new rope of length 20 m.
Is it long enough? You **must** show your working.

(2 marks)

115

Two-way tables

1   70 children each visited a city last week.
The two-way table shows some information
about these visits.

Guided

> Look for rows or columns
> with one empty cell.

	Bath	Warwick	Lichfield	Total
Boys	17 − 7 = ...	14	....	32
Girls	7	....	....	70 − 32 = ...
Total	17	25	....	70

Complete the two-way table.   **(3 marks)**

2   80 children each chose one school activity from dodgeball, football and rounders.
The two-way table shows some information about their choices.

	Dodgeball	Football	Rounders	Total
Girls	12	....	....	41
Boys	....	19	....	....
Total	18	....	25	80

(a) Complete the two-way table.   **(3 marks)**

(b) How many boys chose football?   .... **(1 mark)**

(c) How many girls chose an activity?   .... **(1 mark)**

(d) How many girls chose dodgeball?   .... **(1 mark)**

3   The two-way table shows some information about the colours of motorbikes
and cars in a garage.

	White	Blue	Red	Total
Motorbikes	7	....	....	22
Cars	....	8	....	....
Total	10	17	....	50

(a) Complete the two-way table.   **(3 marks)**

(b) How many motorbikes were there in total?   .... **(1 mark)**

(c) How many cars were there in total?   .... **(1 mark)**

(d) How many cars were not blue?   .... **(1 mark)**

116

Pictograms

1 The pictogram shows the numbers of hours of sunshine in Wolverhampton on Monday, Tuesday and Wednesday of one week.

Monday	○ ○ ○ ○
Tuesday	○ ○ ○
Wednesday	○ ○ ◖
Thursday	
Friday	

○ represents 2 hours

(a) Work out the number of hours of sunshine on Monday.

.... hours   **(1 mark)**

(b) How many more hours of sunshine were there on Monday than Wednesday?

.... hours   **(1 mark)**

There were 4 hours of sunshine on Thursday and 3 hours of sunshine on Friday.

(c) Use this information to complete the pictogram.   **(2 marks)**

2 The pictogram gives information about the number of packets of chocolates sold by a shop some days in one week.

Monday	▭ ▭ ▭ ▭
Tuesday	▭ ▭ ▫
Wednesday	▭ ▭ ▪
Thursday	
Friday	

(a) The total number of packets of chocolates sold on Monday and Tuesday was 130.
Complete the key.

▭ represents .... packets   **(1 mark)**

(b) How many packets of chocolates were sold on Wednesday?

....   **(1 mark)**

70 packets of chocolates were sold on Thursday.
60 packets of chocolates were sold on Friday.

(c) Use this information to complete the pictogram.   **(2 marks)**

Bar charts

1 Shaheen works at an animal shelter for dogs. She has alsatians, bulldogs, labradors and poodles.

Guided

This bar chart shows some information about the alsatians and bulldogs.

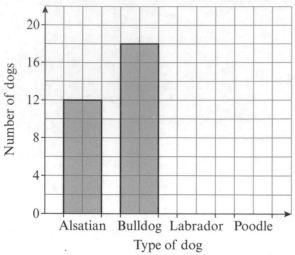

(a) Shaheen also has 8 labradors and 11 poodles in the animal shelter. Complete the bar chart. **(2 marks)**

(b) Which is the most common breed of dog?

.... **(1 mark)**

(c) Work out the total number of dogs in the animal shelter.

12 + ... + ... + ... = ... **(2 marks)**

2

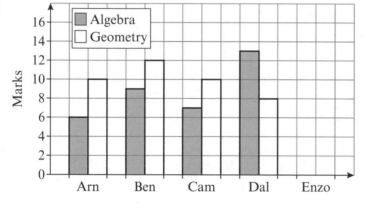

Guided

Some students each sat an algebra test and a geometry test. Each test was out of 15 marks. The dual bar chart shows the results of four of these students.

(a) Who got more marks in their algebra test than their geometry test?

.... **(1 mark)**

(b) How many more marks did Arn get in her geometry test than in her algebra test?

10 − .... = ... **(1 mark)**

Enzo got 9 marks in his algebra test and 14 marks in his geometry test.

(c) Show this information on the dual bar chart. **(2 marks)**

3 Julie asked the students in her class which type of pets they had at home. The bar chart shows some information about the results from her class.

List two things that are wrong with the bar chart.

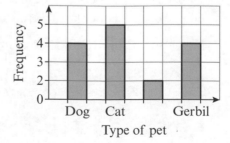

1 ....

2 .... **(2 marks)**

Pie charts

1 Brett carries out a survey of 60 people. He asks them their favourite takeaway. The table shows this information. Draw a pie chart to represent this data.

Guided

Favourite takeaway	Frequency
Indian	14
Chinese	21
Italian	9
Other	16

$\frac{14}{60} \times 360° = $ ....

$\frac{...............}{60} \times 360° = $ ....

$\frac{...............}{60} \times 360° = $ ....

$\frac{...............}{60} \times 360° = $ ....

> You need to calculate the angles first.

(3 marks)

2 Dhruv asked his friends to tell him their favourite colour. The table shows his results. Draw a pie chart to show his results.

Guided

Favourite colour	Frequency
Blue	23
Green	31
Red	22
Yellow	14

$\frac{...............}{90} \times 360° = $ ....

$\frac{...............}{90} \times 360° = $ ....

$\frac{...............}{90} \times 360° = $ ....

$\frac{...............}{90} \times 360° = $ ....

> Draw a line of best fit.

(3 marks)

3 Elaine carries out a survey of some students. The pie chart shows some information about their favourite sport.

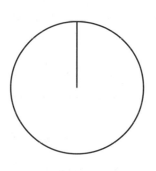

(a) 20 students said that cricket is their favourite sport. How many students said that darts is their favourite sport?

....

(1 mark)

(b) Show that 120 students took part in the survey.

(2 marks)

Scatter graphs

1 The masses of seven magazines and the number of pages in each one were recorded. The scatter graph gives information about these results.

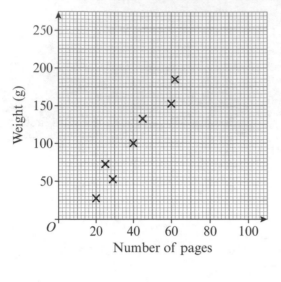

(a) What type of correlation does this scatter graph show?

.... **(2 marks)**

(b) Estimate the mass, in g, of a magazine with 50 pages.

> Draw a line of best fit.

.... g **(2 marks)**

(c) Estimate the mass, in g, of a book with 80 pages.

.... g **(1 mark)**

(d) Nik says, 'As the number of pages increases, the magazines get heavier.' Does the scatter graph support Nik's statement?

.... **(1 mark)**

(e) Make two comments explaining why your estimate in part (c) might not be accurate.

....

.... **(2 marks)**

2 The scatter graph gives information about the price and age of motorbikes.

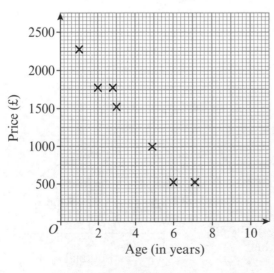

(a) What type of correlation does this scatter graph show?

.... **(2 marks)**

(b) Estimate the price, in £, of a four-year-old motorbike.

£.............................. **(2 marks)**

(c) Comment on the reliability of the estimate in part (b).

....

.... **(1 mark)**

(d) Tim says, 'As the motorbikes get older they get more expensive.' Does the scatter graph support Tim's statement?

.... **(1 mark)**

Averages and range

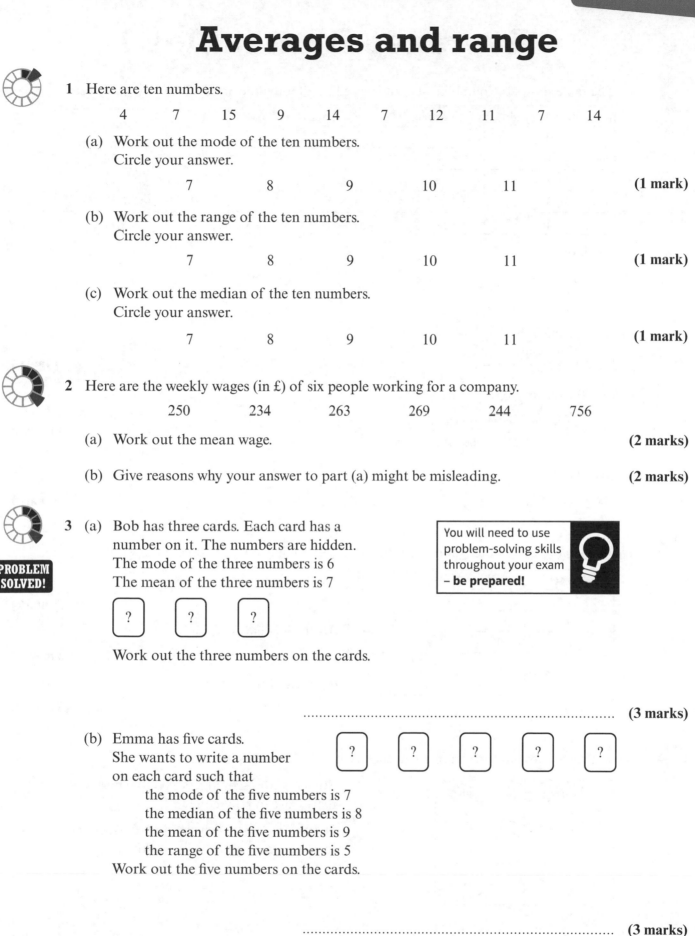

1   Here are ten numbers.

  4   7   15   9   14   7   12   11   7   14

(a) Work out the mode of the ten numbers.
     Circle your answer.

        7   8   9   10   11   **(1 mark)**

(b) Work out the range of the ten numbers.
     Circle your answer.

        7   8   9   10   11   **(1 mark)**

(c) Work out the median of the ten numbers.
     Circle your answer.

        7   8   9   10   11   **(1 mark)**

2   Here are the weekly wages (in £) of six people working for a company.

     250   234   263   269   244   756

(a) Work out the mean wage.   **(2 marks)**

(b) Give reasons why your answer to part (a) might be misleading.   **(2 marks)**

3   (a) Bob has three cards. Each card has a
         number on it. The numbers are hidden.
         The mode of the three numbers is 6
         The mean of the three numbers is 7

         [?]   [?]   [?]

> You will need to use problem-solving skills throughout your exam – **be prepared!**

PROBLEM SOLVED!

         Work out the three numbers on the cards.

         .... **(3 marks)**

(b) Emma has five cards.
     She wants to write a number
     on each card such that

     [?] [?] [?] [?] [?]

          the mode of the five numbers is 7
          the median of the five numbers is 8
          the mean of the five numbers is 9
          the range of the five numbers is 5
     Work out the five numbers on the cards.

          .... **(3 marks)**

Had a go ☐ Nearly there ☐ Nailed it! ☐

Averages from tables 1

Guided

1 The table shows the numbers of goals scored by a football team in each of 30 matches.

Number of goals x	Frequency f	$f \times x$
0	7	$0 \times 7 = \ldots\ldots\ldots$
1	9	$1 \times 9 = \ldots\ldots\ldots$
2	6	$2 \times 6 = \ldots\ldots\ldots$
3	5	$3 \times 5 = \ldots\ldots\ldots$
4	3	$4 \times 3 = \ldots\ldots\ldots$

> To help find the mean:
> Draw an extra column.
> Add up the final column to work out the total number of goals.

Work out

(a) the mode

$\ldots\ldots\ldots$ **(1 mark)**

(b) the median

Median $= \dfrac{30 + 1}{2} = \ldots\ldots\ldots$th value $= \ldots\ldots\ldots$ **(1 mark)**

(c) the mean

Mean $= \dfrac{\text{total number of goals}}{\text{total frequency}} = \dfrac{\ldots\ldots\ldots}{\ldots\ldots\ldots} = \ldots\ldots\ldots$ **(3 marks)**

(d) the range.

Range $=$ highest value $-$ lowest value $= \ldots\ldots\ldots - \ldots\ldots\ldots = \ldots\ldots\ldots$ **(1 mark)**

2 The table shows information about the results of rolling a dice 25 times. Work out

Score	Frequency	
1	3	
2	1	
3	5	
4	3	
5	7	
6	6	

(a) the mode

$\ldots\ldots\ldots\ldots\ldots\ldots\ldots$ **(1 mark)**

(b) the median

$\ldots\ldots\ldots\ldots\ldots\ldots\ldots$ **(2 marks)**

(c) the mean.

$\ldots\ldots\ldots\ldots\ldots\ldots\ldots$ **(3 marks)**

3 The table shows information about the marks of 25 students in a test.

Mark	Frequency
13	3
14	6
15	2
16	6
17	8

Students who scored less than the mean mark have to retake the test.

How many students have to retake the test? You must show your working.

$\ldots\ldots\ldots\ldots\ldots\ldots\ldots$ **(4 marks)**

Averages from tables 2

4 The table shows information about the number of hours spent on the internet last week.

Guided

Number of hours	Frequency f	Midpoint x	f × x
$0 \leqslant h < 2$	6	1	6 × 1 = ...
$2 \leqslant h < 4$	7	3	7 × 3 = ...
$4 \leqslant h < 6$	3	5	3 × 5 = ...
$6 \leqslant h < 8$	9	7	9 × 7 = ...
$8 \leqslant h < 10$	10	9	10 × 9 = ...

> Multiply the frequency by the midpoint of each group.

(a) Which is the modal class?

.... **(1 mark)**

(b) Which class interval which contains the median?

Median $= \dfrac{35 + 1}{2} =$ ....th value.

Median is in class .... **(1 mark)**

(c) Work out an estimate for the mean number of hours.

> Add up the final column to work out the total number of hours.

Estimate of mean $= \dfrac{\text{total number of hours}}{\text{total frequency}} = \dfrac{................}{................}$

Estimate of mean = ... **(4 marks)**

(d) Explain why your answer to part (c) is an estimate.

.... **(1 mark)**

5 Ian asked 25 students how many minutes they each took to get home from school.

Time taken (t minutes)	Frequency
$0 \leqslant t < 10$	6
$10 \leqslant t < 20$	7
$20 \leqslant t < 30$	3
$30 \leqslant t < 40$	9

> Add two columns to the table.

(a) Work out an estimate for the mean time taken. .... **(4 marks)**

(b) Ian realises he has missed out a student. This student takes 32 minutes to get home from school. Ian says, 'The mean time of the students will increase.' Is he correct? Give a reason to support your answer.

.... **(1 mark)**

Line graphs

1 The table shows information about the annual turnover of a company in millions of pounds.

Year	Turnover (£ millions)
2008	13
2009	10
2010	12
2011	13
2012	15
2013	16
2014	18

(a) Draw a time series graph to represent this data.

> Plot the points from the table.

(2 marks)

(b) Describe the trend.

> Use the correct language: upwards or downwards.

....

(1 mark)

2 Joe recorded the number of letters he received each day for a period of time. The graph gives some information about his results.

(a) What is the modal number of letters?

....

(1 mark)

(b) Work out the total number of letters Joe received during this period of time.

(0 × 3) + (1 × ....) + (2 × ....) +

(............ × ....) + (............ × ....)

= ....

(2 marks)

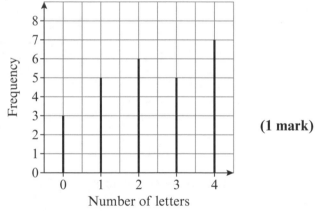

3 The vertical line graph shows the shoe sizes of some children.

(a) What is the modal shoe size.

....

(1 mark)

(b) Work out the mean shoe size.

> You will need to use problem-solving skills throughout your exam – **be prepared!**

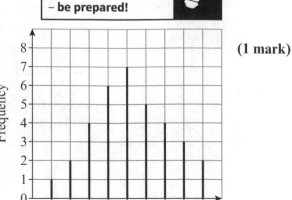

....

(3 marks)

Sampling

PROBLEM SOLVED!

1 Simon wants to find out the number of hours spent on homework in his school each week. He surveys seven children from his class. Here are his results.

You will need to use problem-solving skills throughout your exam – **be prepared!**

9   7   4   2   6   5   3

(a) State one advantage of taking a sample.

.... **(1 mark)**

(b) Use this data to estimate the mean number of hours spent doing homework each week by students.

> Add up all the values and divide by how many there are.

.... **(2 marks)**

(c) Comment on the reliability of your estimate.

.... **(1 mark)**

(d) How could Simon reduce bias in his sample?

.... **(2 marks)**

2 An experiment is carried out by flying paper aeroplanes. The scatter graph shows some information about the distance flown, in m, and the wingspan, in cm. A line of best fit has been drawn.

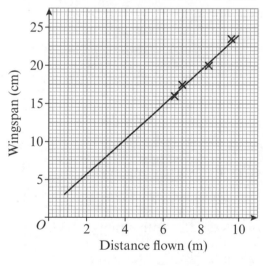

(a) Use the line of best fit to estimate the wingspan of a plane which flies 9 m.

> Draw a vertical line to the line of best fit.

....cm **(1 mark)**

(b) Use the line of best fit to estimate the distance flown by a plane with a wingspan of 5 cm.

.... m **(1 mark)**

(c) Which of your estimates in part (a) or part (b) is more reliable? Give a reason for your answer.

.... **(2 marks)**

(d) Write one way you could improve this experiment to increase the accuracy of your estimates.

.... **(1 mark)**

125

Had a go ☐ **Nearly there** ☐ **Nailed it!** ☐

Comparing data

1 (a) The table shows the results of 2 tests out of 100.

Compare the test scores in Maths and Statistics.

	Mean	Range
Maths	62	14
Statistics	56	20

Students did better in .... because the mean

was ....

Students' results in Maths were less ....

because the range was .... **(2 marks)**

(b) The table shows the amount of rainfall, in mm, in Wolverhampton and Dundee one month.

	Mean	Range
Wolverhampton	25	9
Dundee	39	6

Compare the amount of rainfall in Wolverhampton and Dundee.

....

.... **(2 marks)**

2 Mr Jones kept a record of the number of absences for each student in his class for one term. Here are his results.

1   0   1   8   6   4   3   5   2   3   4   2

(a) Work out the mean.   (b) Work out the range.

.... **(2 marks)**   .... **(1 mark)**

Mr Singh also kept a record of the number of absences in his class.
The mean number of absences was 5 and the range was 5.

(c) Compare the number of absences for each class.

....

.... **(2 marks)**

3 Alex and Joe both play for their local cricket team.
Here are the numbers of runs that Alex scored in his last eight matches:

67   19   4   86   83   9   18   94

In his last 8 matches, Joe's mean number of runs was 45.25 with a range of 38.

(a) Who scored the most runs in their last eight matches?   **(2 marks)**

(b) Who is the more consistent player?
Give reasons for your answers.

(2 marks)

Probability 1

1 On the probability scale, mark with a cross (×) the probability that

(a) you will watch TV tomorrow

├────────┼────────┤
0   1 **(1 mark)**

(b) the sun will not rise tomorrow

├────────┼────────┤
0   1 **(1 mark)**

(c) a coin is tossed and it will land on heads

├────────┼────────┤
0   1 **(1 mark)**

(d) a dice is rolled and it will land on 6.

├────────┼────────┤
0   1 **(1 mark)**

2   impossible   unlikely   evens   likely   certain

Which of these words best describes the likelihood of each of these events?

(a) A dice is rolled and a 7 is shown.

(b) A coin is thrown and lands on tails.

(c) 4 April is the day after 3 April.

.... **(1 mark)**   .... **(1 mark)**   .... **(1 mark)**

3 John rolls an ordinary dice. The faces are labelled 1, 2, 3, 4, 5 and 6.
Work out the probability of rolling

> **Guided**

(a) a 5

....

> How many 5s are there? Remember to write your answer as a fraction.

(1 mark)

(b) an even number

....

> How many even numbers are there?

(1 mark)

(c) a number less than 4

....

(1 mark)

(d) a 10.

....

(1 mark)

4 There are eight discs in a bag.
four discs are red, three discs are green and one disc is blue.
Adam takes a disc at random from the bag.

(a) What is the probability that he takes a red disc?   **(1 mark)**

(b) What is the probability that he takes a blue disc?   **(1 mark)**

(c) What is the probability that he takes a yellow disc?   **(1 mark)**

(d) What is the probability that he takes a disc that is either red or green?   **(1 mark)**

Probability 2

5 A box contains cartons of orange juice, apple juice and mango juice.
The table shows each of the probabilities that a carton of juice taken at
random from the box will be orange or apple.

Guided

Carton of juice	Orange	Apple	Mango
Probability	35%	40%	

> The probabilities have to add up to 100%.

A carton is to be taken at random from the box.
Work out the probability that the carton will be
a mango juice.

100% − (........................... + ...) = ...   **(2 marks)**

6 A spinner can land on A, B, C or D. The table shows information that the spinner
will land on each letter B or C or D.

Letter	A	B	C	D
Probability		0.26	0.36	0.17

The spinner is spun once. Work out the probability that the letter on the spinner

(a) will be A

(b) will **not** be B.   > The probabilities have to add up to 1.

.... **(2 marks)**   ....   **(2 marks)**

PROBLEM SOLVED!

7 Four athletes Andy, Ben, Carl and Daljit take
part in a race. The table shows the probabilities
of Andy or Ben winning the race.

> You will need to use problem-solving skills throughout your exam – **be prepared!**

Athlete	Andy	Ben	Carl	Daljit
Probability	0.3	0.38		x

The probability that Carl will win is three times the probability that Daljit will win.
Work out the probability that the race will be won by Daljit.

....   **(3 marks)**

8 A bag contains counters that are red, blue, green or yellow as shown in the table.

	Red	Blue	Green	Yellow
Number of counters	$3x$	x	$x + 6$	$2x + 6$

The number of red counters is twice the number of green counters.
One counter is chosen at random.
Work out the probability that it is green.   **(4 marks)**

Relative frequency

1   The table shows the total scores when Ethan throws 3 darts 50 times.

Guided

Score	1–30	31–60	61–90	91–120	121–150	151–180
Frequency	14	10	9	8	6	3

He throws another three darts. Estimate the probability that he scores

(a) between 31 and 60

.... out of 50 = $\frac{............}{50}$ **(1 mark)**

(b) more than 90

....   **(2 marks)**

(c) 120 or less.

....   **(2 marks)**

2   A garage keeps records of the cost of repairs it makes to vans. The table gives information about the costs of all repairs which were £500 or less in one month.

Cost (£C)	Frequency
$0 < C \leqslant 100$	20
$100 < C \leqslant 200$	40
$200 < C \leqslant 300$	72
$300 < C \leqslant 400$	30
$400 < C \leqslant 500$	38

(a) Amy needs to repair her van. Estimate the probability her repair costs more than £200

.... **(2 marks)**

(b) Comment on the accuracy of your estimate.

.... **(1 mark)**

3   The diagram shows a six-sided spinner.

The spinner is spun 60 times.
The results are shown in the table.

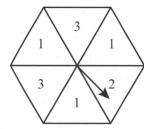

Number	1	2	3
Frequency	34	8	18

(a) Work out the relative frequency of spinning a 2.   **(1 mark)**

(b) Work out the theoretical frequency of spinning a 2
     if the spinner is fair.   **(1 mark)**

(c) Do you think this spinner is fair?
     Give a reason for your answer.   **(1 mark)**

(d) How many times would you expect a 2 to occur if the
     spinner was spun 150 times?   **(1 mark)**

Frequency and outcomes

1 Brett goes to a restaurant.
He can choose from three types of curry and
three types of naan.
Brett is going to choose one curry and one naan.
Work out the probability that he chooses a
lamb curry with a butter naan.

> Guided

Curry	Naan
Chicken	Plain
Lamb	Garlic
Vegetable	Butter

Label chicken as C, and so on.

(C, ...) (C, ...) (C, ...)

(L, ...) (L, ...) (L, ...)

(V, ...) (V, ...) (V, ...)

Probability = ... **(3 marks)**

2 Martin is holding three cards, labelled X, Y and Z.
He mixes them up and then asks Neil to choose a card at random.

(a) What is the probability that Neil chooses card Y?

.... **(1 mark)**

Neil replaces his card and Martin mixes the cards up again.
Martin then asks Tej to choose a card.

(b) Complete the table of possible outcomes.

Neil's card									
Tej's card									

(2 marks)

(c) Work out the probability that Neil and Tej both choose the same card.

.... **(1 mark)**

(d) Work out the probability that Neil and Tej choose different cards.

.... **(1 mark)**

3 120 adults were asked if they voted in the general election. 58 of these adults were male.
Seven of the females did not vote. 103 of the adults voted.

(a) Complete this frequency tree
to show this information.

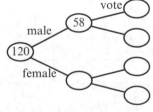

Each frequency is equal to the sum of its branches.

(3 marks)

One of the males is chosen at random.

(b) Work out the probability that this male did not vote.

.... **(2 marks)**

4 A bag contains 50 counters. They are all either green or blue.
A counter is chosen at random. The probability that it is green is $\frac{3}{10}$
Work out the number of blue counters in the bag.

.... **(2 marks)**

Venn diagrams

1 These diagrams represent the subjects studied at college by a group of students. There are 30 students in total in each case. For each diagram

Guided

(a) work out the value of x

(b) state which set x represents.

(i)
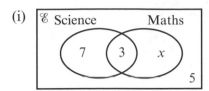

30 − (.......................... + ... + ...)

x = ...   **(1 mark)**

Students who only study ....   **(1 mark)**

(ii)
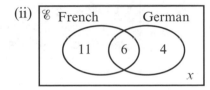

(iii)
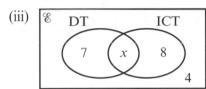

x = ...   **(1 mark)** x = ...   **(1 mark)**

....   **(1 mark)**   ....   **(1 mark)**

2 The Venn diagram shows information about musical instruments played by 40 students. A student is chosen at random. Work out the probability that this student

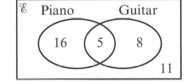

(a) plays the piano and   (b) plays neither   (c) plays the piano.
     the guitar   instrument

.... **(1 mark)**   .... **(1 mark)**   .... **(1 mark)**

3 $\mathscr{E}$ = {1, 2, 3, 4, 5, 6 ,7, 8, 9, 10, 11, 12}

A = odd numbers
B = factors of 12

(a) Complete the Venn diagram.   **(3 marks)**

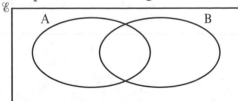

(b) One of the numbers is chosen at random.
     Work out P(A ∩ B).   **(1 mark)**

Independent events

Guided

PROBLEM SOLVED!

1 Marcus has ten counters in a bag.
Three of the counters are yellow and the remaining counters are blue.
Marcus chooses a counter at random and notes the colour.
He then puts the counter back into the bag.
He chooses another counter at random and notes the colour.
Work out the probability that

> You will need to use problem-solving skills throughout your exam – **be prepared!**

(a) both counters will be yellow

$$\frac{3}{10} \times \frac{\text{............}}{10} = \text{................}$$ **(2 marks)**

(b) both counters will be blue

$$\frac{\text{............}}{10} \times \frac{\text{............}}{10} = \text{................}$$ **(2 marks)**

(c) the counters will be different colours.

$$\frac{\text{............}}{\text{............}} \times \frac{\text{............}}{\text{............}} + \frac{\text{............}}{\text{............}} \times \frac{\text{............}}{\text{............}} = \text{............}$$ **(3 marks)**

> 1st Yellow , 2nd Blue and 1st Blue, 2nd Yellow are different outcomes.

2 A bag contains three blue marbles and nine red marbles.
A marble is chosen at random, replaced, and then another is taken out.

Guided

(a) Complete the probability tree diagram.

First marble   Second marble

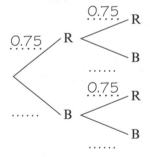

0.75 ⟋ R
0.75 ⟋ R ⟨
            .... ⟍ B
.... ⟍ B ⟨ 0.75 ⟋ R
            .... ⟍ B

(2 marks)

(b) Work out the probability that both marbles are the same colour.

.... **(3 marks)**

3 Nav and Asha each take a motorcycle test.
The probability that Nav will pass is 0.9
The probability that Asha will pass is 0.8

(a) Complete the probability tree diagram.

Nav   Asha

.... ⟋ Pass
.... ⟍ Fail

(2 marks)

(b) Work out the probability that only one of them will pass the test.

.... **(3 marks)**

Problem-solving practice 1

1 Here are four boxes, A, B, C and D.
They contain balls coloured red (R), green (G) and yellow (Y).

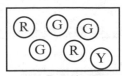 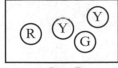 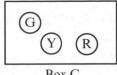 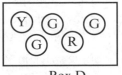

Box A   Box B   Box C   Box D

A ball is picked at random from each box.

(a) Which box gives the **greatest** chance of picking a green ball?
You **must** show your working.   **(2 marks)**

(b) Which two boxes give the **same** chance of picking a red ball?   **(1 mark)**

2 80 students each study 1 of 3 languages.
The two-way table shows some information about these students.

	French	German	Spanish	Total
Female	15	....	....	39
Male	....	17	....	41
Total	31	28	....	80

(a) Complete the two-way table.   **(3 marks)**

(b) One of these students is to be picked at random.
What is the probability that this student studies French?

.... **(1 mark)**

3 In a class of 25 students, 8 study Latin, 10 study Mandarin and 3 study both.

(a) Draw a Venn diagram to represent this information.   **(3 marks)**

A student is chosen at random.
Work out the probability that the student

(b) does not study Latin and does study Mandarin

.... **(2 marks)**

(c) studies Latin or Mandarin but not both.

.... **(2 marks)**

Problem-solving practice 2

4 Heather carries out a survey of 180 Year 11 students.
She asks them their favourite snack.
She draws this accurate pie chart.
Use the pie chart to complete the table.

Favourite snack in Year 11

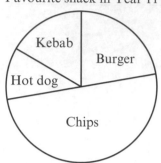

Diagram accurately drawn

Favourite snack in Year 11	Frequency	Angle
Burger	40	....
Chips	90	180°
Hot dog	....	....
Kebab	....	....
Total	180	

(4 marks)

5 A school snack bar offers a choice of four snacks. The four snacks are burgers, wraps, fruit and salad. Students can choose one of these four snacks.
The table shows the probability that a student will choose a burger or a wrap.

Snack	Burger	Wrap	Fruit	Salad
Probability	0.25	0.15		

The probability that a student chooses fruit is twice the probability that a student chooses salad.
One student is chosen at random from the students who use the snack bar.

(a) Work out the probability that the student did not choose a burger.   .... **(2 marks)**

(b) Work out the probability that the student chose salad.   .... **(2 marks)**

(c) 200 students used the snack bar on Tuesday.
Estimate the number of students who chose a wrap.

.... **(2 marks)**

6 Jamie and Rajiv each take an entrance exam.
The probability that Jamie will pass the entrance exam is 0.7
The probability that Rajiv will pass the entrance exam is 0.75

(a) Complete the probability tree diagram.

(b) Work out the probability that only one of them will pass the entrance exam.

Jamie   Rajiv

...... ⟋ Pass

...... ⟍ Fail

(2 marks)   .... **(3 marks)**

Paper 1

Practice exam paper

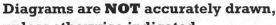
Foundation Tier
Time: 1 hour 30 minutes
Calculators may be used.
Diagrams are NOT accurately drawn,
unless otherwise indicated.
You must show all your working out.

1. The diagram represents a straight road that joins four villages.

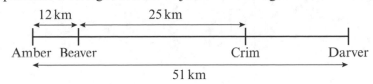

   (a) Work out the distance from Crim to Darver.

   .... **(1 mark)**

   Lewis walks from Amber to Crim.
   Harry walks from Beaver to Darver.

   (b) Who walks the furthest?
       Give a reason for your answer. ....

   .... **(2 marks)**

2. Circle the fraction that is not equivalent to $\frac{5}{6}$ **(1 mark)**

   $\dfrac{10}{12}$ $\dfrac{15}{18}$ $\dfrac{20}{25}$ $\dfrac{25}{30}$

3. One of the letters in the word DIAGRAMMATIC is selected at random.
   Circle the probability that it is the letter A. **(1 mark)**

   $\dfrac{1}{6}$ $\dfrac{1}{4}$ $\dfrac{1}{12}$ $\dfrac{1}{3}$

4. Which of these numbers is both a square number and a cube number?
   Circle your answer. **(1 mark)**

   27   36   64   100

5. Sandeep recorded the temperature at different times on 1 January 2016.

Time of day	Temperature (°C)
3 am	−11
7 am	−5
Noon	7
4 pm	5
8 pm	−3
Midnight	−9

   What is

   (a) the highest temperature .... **(1 mark)**

   (b) the lowest temperature? .... **(1 mark)**

   Work out the difference in the temperature between

   (c) 3 am and 7 am .... **(1 mark)**

   (d) noon and 8 pm. .... **(1 mark)**

6. (a) Insert **one** pair of brackets to make this calculation correct. **(1 mark)**
   $7 \times 8 - 5 + 3 = 24$

   (b) Insert **two** pairs of brackets to make this calculation correct. **(1 mark)**
   $9 + 1 \times 8 \div 4 - 2 = 40$

7. How many millimetres are there in 0.26 metres?
   Circle your answer. **(1 mark)**

   | 2.6 | 26 | 260 | 2600 |

8. Simplify
   (a) $5x + 8x - 3x$ .... **(1 mark)**
   (b) $8e - 6f + 7e + 8e - 2f + 4$ .... **(2 marks)**

9. PQR is a straight line.
   $SR = QR$

   Work out the value of x
   Give reasons for your answer.

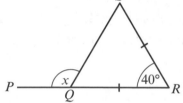

.... **(3 marks)**

10. The diagram shows part of a map.
    It shows the positions of a
    statue and a cathedral.

    The scale of the map is 1:10 000
    Work out the real distance
    between the statue and the
    cathedral.
    Give your answer in metres.

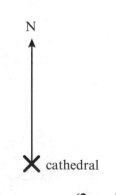

.... **(2 marks)**

11. 60 students were asked to name their favourite sport. Here are the results.

Basketball	Football	Rugby	Hockey
12	25	15	8

Draw a pie chart to represent this data. **(4 marks)**

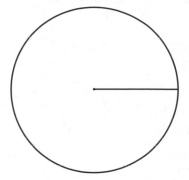

12. A school has 420 students.
65% of the students eat school lunch.
$\frac{3}{20}$ of the students take a packed lunch.

The rest go home for lunch.
How many students go home for lunch?
You **must** show your working. **(3 marks)**

13. Factorise

(a) $5x + 10$ .... **(1 mark)**

(b) $x^2 - 6x$ .... **(1 mark)**

14. A bag contains only red, white and blue counters.
One counter is chosen at random.
The probability of selecting a red counter is 0.45
The probability of selecting a white counter is 0.25
Circle the probability of selecting a blue counter. **(1 mark)**

     0.25   0.3   0.33   0.4

15. George wants to buy a laptop for £634
He has already saved £115
Each week his pay is £120
He saves 35% of his pay.
How many **more** weeks must he save? **(4 marks)**

16. There are 40 students in a class.
16 of the students study Latin.
19 of the students study Spanish.
Seven of the students study both Latin and Spanish.

(a) Draw a Venn diagram to represent this information.

(3 marks)

(b) Show that 30% of the students do **not** study Latin or Spanish.

(2 marks)

17. Lucy drove a distance of 120 km in 1 hour 20 minutes.

What was her average speed in km/hour?   **(3 marks)**

18. Amy shares a bag of sweets with her friends.

She gives Beth $\frac{2}{5}$ of the sweets.

She gives Carl $\frac{3}{10}$ of the sweets.

She has 12 sweets left.
How many sweets does Amy give to Beth?   .... **(4 marks)**

19. In England, in 2016, female cattle aged two years or more were either beef cattle or dairy cattle.
The ratio of beef cattle to dairy cattle was $1:1\frac{1}{2}$

(a) What fraction of the female cattle were dairy cattle?   **(2 marks)**

(b) There were 900 female beef cattle.

What was the total number of female cattle?   **(2 marks)**

20. Here are two number machines.

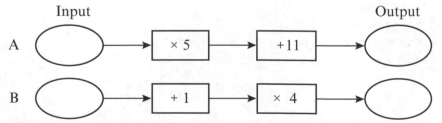

Both machines have the same input.
Work out the input that makes the output of B twice the output of A.   **(4 marks)**

21. In a sale a TV was sold for two-thirds of its original price.
The sale price was £306
What was the original price of the TV?   **(3 marks)**

22. Here are the first four terms of an arithmetic sequence.

$$7 \qquad 11 \qquad 15 \qquad 19$$

Work out an expression for the nth term of this sequence. **(2 marks)**

23. The diagram shows a quadrant of a circle of radius 12 cm.

12 cm

Work out the area of the quadrant.
Give your answer in terms of π. **(2 marks)**

24. The table shows how many hours of homework, per week, were done by 30 students.

Time (t hours)	Number of students
$0 \leqslant t < 2$	2
$2 \leqslant t < 4$	3
$4 \leqslant t < 6$	4
$6 \leqslant t < 8$	12
$8 \leqslant t < 10$	5
$10 \leqslant t < 12$	4

Work out an estimate for the mean number of hours of homework done per week. **(4 marks)**

25. Which sequence is a geometric progression? **(1 mark)**
Circle the answer.

$$1 \quad 2 \quad 4 \quad 6 \qquad\qquad 1 \quad 8 \quad 16 \quad 32$$

$$1 \quad 4 \quad 16 \quad 64 \qquad\qquad 1 \quad 4 \quad 8 \quad 12$$

26. (a) Write 50 000 in standard form.

.... **(1 mark)**

(b) Write 9.6×10^{-5} as an ordinary number.

.... **(1 mark)**

(c) Work out the value of $(5 \times 10^4) \times (3 \times 10^6)$
Give your answer in standard form.

.... **(2 marks)**

27. Shapes *ABCD* and *PQRS* are similar.

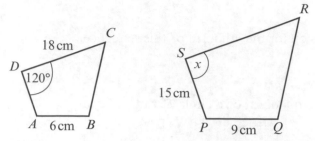

(a) What is the size of the angle marked *x*? .... **(1 mark)**

(b) Work out the length of *AD*

.... **(2 marks)**

(c) Work out the length of *RS*

.... **(2 marks)**

28. (a) Complete the table of values for $y = \dfrac{8}{x}$

x	0.5	1	2	4	5	8
y		8			1.6	

(2 marks)

(b) Draw the graph of $y = \dfrac{8}{x}$

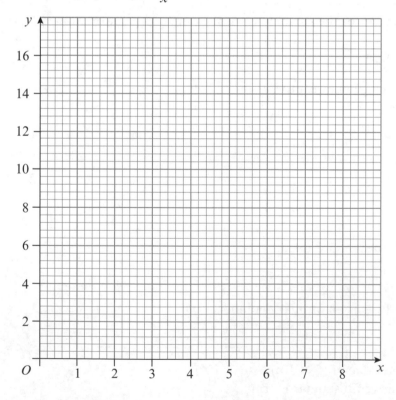

(2 marks)

TOTAL MARKS FOR PAPER = 80 MARKS

Answers

NUMBER

1. Place value
1 (a) 9351
   (b) Four thousand, one hundred and ninety-six
   (c) 5000

2

	ten thousands	thousands	hundreds	tens	units
	1	2	0	6	0

3 (a) 49, 127, 146, 165, 169
   (b) 7028, 7249, 7429, 7924, 7942
4 (a) 347, 374, 437, 473, 734, 743
   (b) 347, 374, 437, 473, 734, 743
5 8614
6 110 paper plates

2. Negative numbers
1 $-11, -4, 0, 4, 6$
2 (a) -2 (b) -3 (c) -10 (d) -16
3 (a) $-2°C$ (b) $-7°C$ (c) $-15°C$ (d) $22°C$
4 (a) $50°C$
   (b) New Delhi
   (c) No, answer is $-9°C$.
5 (i) -14 (ii) -7 (iii) 24 (iv) 7

3. Rounding numbers
1 (a) 27 000   (b) 6500   (c) 87 540
2 (a) 8.6   (b) 8.636   (c) 8.64
3 (a) 0.003   (b) 0.003 5   (c) 0.003 47
4 (a) 40 000   (b) 39 000   (c) 38 700
5 (a) 20.4 g   (b) 170 g   (c) 300
6 No, correct answer is 0.0235

4. Adding and subtracting
1 (a) 1023   (b) 577
2 (a) 8178   (b) 177
3 41
4 £5.80
5 He is not correct. Coffee costs £2.06

5. Multiplying and dividing
1 (a) 1909   (b) 243
2 84
3 756
4 (a) 43 290   (b) 34
5 (a) 5 boxes   (b) 9 boxes
6 (a) 125 chocolates   (b) 19 chocolates

6. Decimals and place value
1 (a) Seven tenths or $\frac{7}{10}$
   (b) Eight hundredths or $\frac{8}{100}$
   (c) Four thousandths or $\frac{4}{1000}$
2 1.4, 3.2, 6.2, 6.4, 12.8
3 0.05, 0.6, 0.61, 0.611, 0.613
4 0.7, 0.725, 0.73, 0.778, 0.78
5 (a) 2451   (b) 24.51   (c) 4.3
6 He is incorrect because $435.2 \div 13.6 = 32$

7. Operations on decimals
1 £282.60
2 (a) 14.63   (b) 75.36   (c) 117.12
   (d) 0.0329   (e) 13.9   (f) 63
3 £15.40
4 £20.94

8. Squares, cubes and roots
1 (a) 16   (b) 8   (c) 9
   (d) 8   (e) 4   (f) -5
2 (a) 81   (b) 125   (c) 12   (d) 6
3 52
4 89
5 (a) 36 or 49   (b) 8   (c) 49
6 No because $2 \times 2 \times 2 = 8$.
7 No because $16 + 4 + 1 = 21$ which is odd.

9. Indices
1 (a) 4^2 (b) 4^5
2 (a) 5^9 (b) 5^3 (c) 5^4 (d) 5^{12}
3 4^{10}
4 (a) 9^{-1} (b) 9^{-4}
5 (a) 8^3 (b) 8^2 (c) 8^8 (d) 8^{20}
6 (a) 1   (b) $\frac{1}{49}$ (c) $\frac{1}{64}$ (d) $\frac{27}{64}$
   (e) $\frac{27}{64}$ (f) $\frac{25}{16}$
7 $x = 8$

10. Estimation
1 (a) 14 000   (b) 6
2 125
3 17 500
4 750
5 (a) 240
   (b) Harry used 0.5 as $\frac{1}{2}$ and divided by 2.
6 (a) 432 cm²   (b) underestimate

11. Factors, multiples and primes
1 (a) $1 \times 36, 2 \times 18, 3 \times 12, 4 \times 9, 6 \times 6$
   (b) 7, 14, 21, 28, 35, 42, 49, 56, 63, 70
2 (a) 2 or 6   (b) 21 or 49   (c) 6 and 8
3 $14 + 7 + 1$ or $14 + 7 + 2$
4 25
5 41, 43, 47
6 (a) 2×3^3 (b) $2^5 \times 3$ (c) $2 \times 3^2 \times 7$ (d) $2^2 \times 3^2 \times 7$

12. HCF and LCM
1 (a) 12   (b) 60
2 (a) (i) $2 \times 3^2 \times 5$ (ii) $2 \times 3 \times 5 \times 7$
   (b) 30   (c) 630
3 (a) 12   (b) 144

13. Fractions

1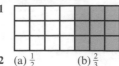

2 (a) $\frac{1}{2}$ (b) $\frac{2}{3}$ (c) $\frac{7}{24}$ (d) $\frac{2}{7}$

3 (a) $\frac{3}{4}$ (b) $\frac{3}{4}$

4 (a) £45   (b) £64   (c) £140   (d) £150

5 $\frac{7}{20}$

6 £66

14. Operations on fractions

1 (a) $\frac{11}{15}$ (b) $\frac{11}{20}$ (c) $\frac{69}{56}$ (d) $-\frac{1}{63}$

2 (a) $\frac{1}{6}$ (b) $\frac{15}{44}$ (c) $\frac{40}{15}$ (d) $\frac{18}{12}$

3 $\frac{4}{15}$

4 (a) $\frac{8}{35}$ (b) 48 litres

15. Mixed numbers

1 (a) $6\frac{11}{20}$ (b) $2\frac{1}{10}$

2 (a) $\frac{23}{6}$ (b) $\frac{10}{3}$

3 (a) 9   (b) $\frac{48}{13}$

4 $\frac{71}{12}$ hours

5 $6 \times 1\frac{4}{5} = 6 \times \frac{9}{5} = 54 \div 5 = 10\frac{4}{5}$, so no he doesn't have enough

16. Calculator and number skills

1 (a) 15   (b) 95   (c) 18   (d) 81

2 (a) 6   (b) 1

3 5

4 1.751 592 357

5 (a) 8.5625   (b) 9

6 (a) 1.248 005 424   (b) 1.25

7 (a) 2.798 083 024   (b) 2.8

17. Standard form 1

1 (a) 4.5×10^4 (b) 0.000 034   (c) 2.8×10^7

2 (a) 5.67×10^5 (b) 5.67×10^{-5} (c) 5.67×10^{10}

3 (a) 6.74×10^6 (b) 7.3×10^6 (c) 6.2×10^6

4 7.5×10^7

5 4×10^2

6 5.2×10^4 km/h

18. Standard form 2

7 (a) 1.8×10^4 (b) 2×10^{20}

8 (a) 7.01×10^4 (b) 7.52×10^5

9 (a) 1.12×10^{10} (b) 4.48×10^8 (c) 7×10^{10}

10 1.44×10^8

11 15 000

19. Counting strategies

1 (A,2), (A,3), (R,1), (R,2), (R,3), (T,1), (T,2), (T,3)

2 (C,P), (C,G), (C,B), (L,P), (L,G), (L,B), (V,P), (V,G), (V,B)

3 (W,X), (W,Y), (W,Z), (X,Y), (X,Z), (Y,Z)

4 6

5 6

6 10

20. Problem-solving practice 1

1 2, 5, 13, 17 (other possibilities are: 2 + 3 + 7 + 23, 2 + 3 + 11 + 19, 2 + 3 + 13 + 17, 2 + 5 + 11 + 17)

2 £7

3 (a) odd and prime
  (b) 'No', cannot be both prime and square
  (c) e.g. 25: odd, square and two-digit; 23: odd, prime and two-digit (and many others)

4 $\frac{5}{12}$

5 60 boxes

6 6

21. Problem-solving practice 2

7 A = 31, B = 23, C = 40, D = 20

8 18 cm

9 9 am

10 $(4.86 \times 10^{-5}) \times (6.2 \times 10^4) = 3.01 \approx 3$ m

ALGEBRA

22. Collecting like terms

1 (a) $5x$ (b) $6w$

2 (a) $4x + 2y$ (b) $5ab$ (c) $t + 11v$ (d) $3c - 2d$

3 (a) $2x$ (b) $3t^2$ (c) $3a + 8b + 7$
  (d) $x - 7y$ (e) $6p + q$ (f) $2a + 6b + 5$

4 $7m - 3n$

5 (a) expression   (b) formula   (c) equation

23. Simplifying expressions

1 (a) y^2 (b) $3mt$

2 (a) w^4 (b) $28d$ (c) $30k$ (d) $40jk$

3 (a) $15x^2$ (b) $6ef$ (c) $4a$ (d) $8b$
  (e) $6p + q$ (f) $2a + 6b + 5$

4 (a) $35gh$ (b) $8t^3$ (c) $3x$ (d) $4y$

5 (a) $24abc$ (b) $8p$

6 $2(x \times y)$ and $4xy \div 2$

24. Algebraic indices

1 (a) a^9 (b) a^3 (c) a^4 (d) a^{12}

2 (a) t^3 (b) t^3 (c) t^8 (d) t^6

3 (a) x^{12} (b) $64x^6$ (c) $8x^9$ (d) $12x^7$
  (e) $12x^7y^5$ (f) $3x^2y^3$

4 (a) 8   (b) 5   (c) 5

5 3.5

25. Substitution

1 (a) 23   (b) −11

2 (a) 34   (b) 19

3 xy

4 (a) −6   (b) 100   (c) 12

5 (a) 104   (b) 23   (c) 20

6 Abbie is correct, $\frac{1}{2} \times 2 \times 3^2 = 9$

26. Formulae

1 1 hour 45 minutes

2 £200

3 30

4 36

5 18

6 52

7 (a) −4°F   (b) −40°

27. Writing formulae

1. $4g + 5h$
2. $S = 10m + 20n$
3. $P = 30n + 50$
4. $T = 4 + 3x$
5. $P = 10x + 2$
6. (a) $B = n + 4$
   (b) Yes, because Carl is $3n$ years old.

28. Expanding brackets

1. (a) $3x + 6$ (b) $4x + 20$ (c) $5x - 15$
   (d) $12x + 18$ (e) $5 - 20x$ (f) $21x - 56$
2. (a) $-3x + 9$ (b) $-4x - 12$ (c) $-6x + 30$
   (d) $-4x - 6$ (e) $-8x + 2$ (f) $-2x + 4$
3. (a) $x^2 + x$ (b) $x^2 + 5x$ (c) $2x^2 - 18x$
   (d) $6x^2 - 9x$ (e) $-2x^2 + 3x$ (f) $-12x^2 + 15x$
4. (a) $7x + 6$ (b) $5x + 14$ (c) $13x - 11$
   (d) $6x^2 - 20x$
5. No he is not correct. Sign error in second expansion.
   Correct answer is $18x + 3x^2$

29. Factorising

1. (a) $3(x + 2)$ (b) $6(a + 3)$ (c) $2(p - 3)$
   (d) $5(y - 3)$ (e) $3(t + 8)$ (f) $4(g - 5)$
2. (a) $x(x + 6)$ (b) $x(x - 4)$ (c) $x(x - 9)$
   (d) $x(x - 12)$ (e) $x(x + 5)$ (f) $x(x - 1)$
3. (a) $3p(p + 2)$ (b) $8y(y - 3)$ (c) $9t(t - 4)$
   (d) $4d(d + 3)$ (e) $6x(x - 3)$ (f) $7n(n - 5)$
4. $2x^2 - 3xy, 3, 3x, 2x - 3y$
5. A4, B3, C1, D2

30. Linear equations 1

1. (a) 16   (b) -5 (c) -15
   (d) -36 (e) -120 (f) -9
2. (a) 5   (b) 12   (c) 2
   (d) 3   (e) 19   (f) -46
3. (a) 6   (b) -3 (c) -4
   (d) -4 (e) 15   (f) -24
4. $x + 4x + x - 6 = 48$ giving $x = 9$
   So the numbers are 9, 36 and 3.

31. Linear equations 2

5. (a) 2   (b) -8 (c) $\frac{3}{4}$
6. (a) 5   (b) 2   (c) -3 (d) 3
7. No she is not correct.
   $12x - 15 = 2x + 9$
   $12x - 2x = 9 + 15$
   $10x = 24$
   $x = 2.4$
   Or:
   $3(4x - 5) = 3(4 \times 0.4 - 5) = -10.2$
   $2x + 9 = 2 \times 0.4 + 9 = 9.8$
   When $x = 0.4$, $3(4x - 5) \neq 2x + 9$

32. Number machines

1. $x = -2$
2. (a) $\times 5$ and -18 (b) $n = 4.5$
3. -1.5

33. Inequalities

1. (a) $x \leqslant 4$ (b) $x > -1$
   (c) $-2 < x < 4$ (d) $-1 < x \leqslant 5$
2. (a)

   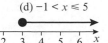

   (b)

   (c)

   (d)

3. (a) $x = -2, -1, 0, 1, 2, 3, 4$ (b) $x = -2, -1, 0, 1, 2$
   (c) $x = -3, -2, -1, 0, 1$
4. (a) $51 - 34 = 17$ (b) $44 + 28 = 72$

34. Solving inequalities

1. (a) $x \leqslant 10$ (b) $x > 5$ (c) $x \geqslant 4$
   (d) $x \leqslant -\frac{16}{3}$ (e) $x > 2$ (f) $x \geqslant -\frac{2}{3}$
2. (a) $x \geqslant 6$ (b) $x < 7$ (c) $x > 3$
   (d) $x \leqslant -3$
3. (a) $x = -2, -1, 0, 1$
   (b) $x = -1, 0, 1, 2, 3, 4$
4. $x = 5$
5. $70 < x + 3x + x + 7 < 100$
   $63 < 5x < 93$, or $12.6 < x < 18.6$
   so $x = 16$, the only square number in this range.
   So, the numbers are 16, 48 and 23.

35. Sequences 1

1. (a) [ ][ ][ ][ ]   (b) 16
   (c) No, because the pattern will never be divisible by 3.
2. (a) 18, 22   (b) 23, 28   (c) 81, 243   (d) 25, 36
3. 1, 3, 4, 7, 11, 18
4. (a) 22, 18
   (b) Ravina is incorrect. All terms end in even digits.
5. 32

36. Sequences 2

6. (a) $4n + 1$ (b) $3n - 1$ (c) $7n - 5$ (d) $5n + 3$
7. $3n + 1$
8. (a) $4n - 1$
   (b) $4n - 1 = 199$
   $4n = 200$
   $n = 50$
   n is an integer therefore 199 is part of the sequence.
9. 33, 45, 57

37. Coordinates

1. (a) (4, 3)   (b) (−2, −3)
   (c) (d)

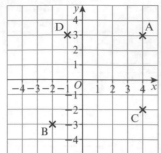

   (e) Kite
2. (a) (5, 9)   (b) $(5, \frac{11}{2})$ (c) (−2, 10)   (d) (−4, 3)
3. (a) $(\frac{13}{2}, 10)$ (b) $(\frac{11}{2}, \frac{17}{2})$
4. B (8, 6) and then P (2, 10)

38. Gradients of lines

1. (a) $\frac{3}{2}$ (b) 2
2. (a) $-\frac{8}{5}$ (b) $-\frac{3}{5}$
3. (a) $\frac{20}{8} = 2.5$ cm/s

39. Straight-line graphs 1

1. (a) −5, −3, −1, 1, 3, 5
   (b)

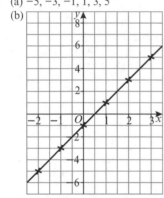

2.

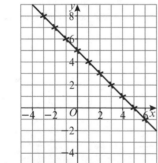

3. $y = 2x + 4$

40. Straight-line graphs 2

4. $y = 6x + 3$
5. (a) $y = 3x - 1$ (b) $y = -2x + 12$
   (c) $y = 4x + 15$ (d) $y = 4x - 2$
6. (a) $y = 2x - 4$ (b) $y = -3x + 2$
7. $y = 4x + 5$

41. Real-life graphs

1. (a) 96 km   (b) Marseille
2. (a) 4.6 m   (b) Sandeep
3. (a) £60   (b) Yes, gradient = 1.5

42. Distance–time graphs

1. (a) 10:00 am   (b) 6 km
   (c) 15 minutes   (d) 18 km/h
2. (a) 30 minutes   (b) 2 km
   (c)

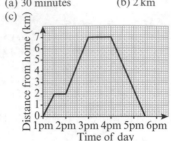

3. (a)

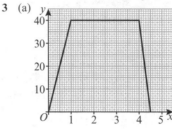

   (b) 80 km/h

43. Rates of change

1.

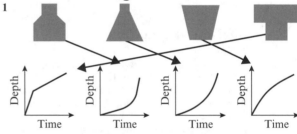

2. (a) £2400   (b) 300
   (c) For every month, Dan saves £300.
3. (a) 3 m/s^2 (b) The velocity is constant at 30 m/s.
   (c) −1.5 m/s^2

44. Expanding double brackets

1. (a) $x^2 + 8x + 15$
   (b) $x^2 + 5x + 6$
   (c) $x^2 + 5x + 4$
   (d) $x^2 - 3x - 10$
   (e) $x^2 - 7x + 10$
2. (a) $15x^2 - 23x + 4$
   (b) $16x^2 - 14x + 3$
   (c) $28x^2 - 55x + 25$
   (d) $x^2 + 6x + 9$
   (e) $4x^2 - 20x + 25$
3. Area of ABCH $= (x + 2)(x + 3) = x^2 + 5x + 6$
   Area of HCFG $= (x + 2)(x + 2) = x^2 + 4x + 4$
   Area of CDEF $= 4(x + 2) = 4x + 8$
   Total area $=$
   $x^2 + 5x + 6 + x^2 + 4x + 4 + 4x + 8 = 2x^2 + 13x + 18$

45. Quadratic graphs

1 (a)

x	-3	-2	-1	0	1	2	3
y	7	2	-1	-2	-1	2	7

(b)
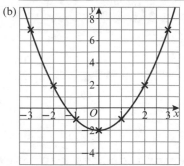

(c) $(0, -2)$ (d) 4.2 (or 4.3)

2 (a)

x	-1	0	1	2	3	4	5
y	8	3	0	-1	0	3	8

(b)
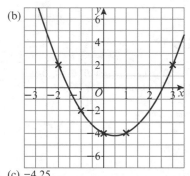

(c) $(2, -1)$

46. Using quadratic graphs

1 (a)

x	-3	-2	-1	0	1	2	3
y	8	2	-2	-4	-4	-2	2

(b)
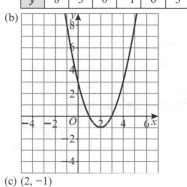

(c) -4.25
(d) $-1.6, 2.6$

2 (a)

x	-1	0	1	2	3	4	5
y	-3	2	5	6	5	2	-3

(b)
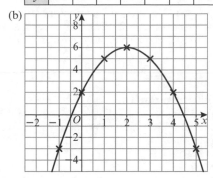

(c) $(2, 6)$
(d) -0.4 and 4.4

47. Factorising quadratics

1 (a) $(x + 1)(x + 3)$ (b) $(x + 10)(x + 1)$
   (c) $(x + 5)(x + 1)$ (d) $(x - 10)(x - 1)$
   (e) $(x - 2)(x - 10)$ (f) $(x - 7)(x - 2)$
2 (a) $(x + 7)(x - 1)$ (b) $(x + 5)(x - 1)$
   (c) $(x - 5)(x + 3)$
3 (a) $(x - 11)(x - 2)$ (b) $(x - 8)(x + 2)$
   (c) $(x - 4)(x - 10)$
4 (a) $(x - 3)(x + 3)$ (b) $(x - 12)(x + 12)$
   (c) $(x - 9)(x + 9)$ (d) $(x - 8)(x + 8)$
   (e) $(x - 1)(x + 1)$ (f) $(x - 13)(x + 13)$

48. Quadratic equations

1 (a) $x = 0, x = 3$ (b) $x = 0, x = -5$
   (c) $x = 0, x = 7$
2 (a) $x = -4, x = -2$ (b) $x = 4, x = 3$
   (c) $x = -5, x = -4$ (d) $x = -1, x = -7$
   (e) $x = 6, x = -4$
3 (a) $x = -2, x = 2$ (b) $x = -5, x = 5$
   (c) $x = -7, x = 7$ (d) $x = -11, x = 11$
   (e) $x = -3, x = 3$
4 $(x - 2)(x + 7) = 0$

49. Cubic and reciprocal graphs

1 (a)

x	-2	-1	0	1	2	3
y	-2	1	-2	-5	-2	13

(b)
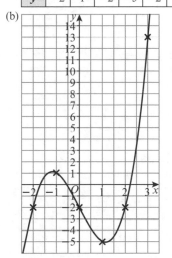

(c) $x \approx 2.6$
(d) $x \approx -1.7, -0.5$ and 2.2
2 (a) B   (b) D   (c) A   (d) C

50. Simultaneous equations

1 (a) $x = 1.5, y = 2.5$ (b) $x = 5, y = -2$
2 (a)
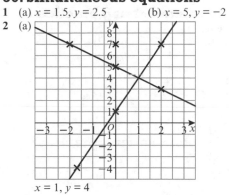

$x = 1, y = 4$

(b)

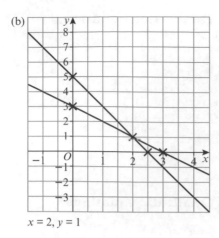

$x = 2, y = 1$

51. Rearranging formulae
1   $x = 3$
2   $t = 2.5$ seconds
3   (a) $t = \frac{1}{10}(v - u)$ (b) $n = \frac{1}{6}(m - 19)$

    (c) $u = \frac{1}{t}(d - at^2)$ (d) $D = \frac{1}{6}(A - P)$

4   (a) $t = \frac{d}{s}$ (b) $h = \frac{4d^2}{5}$

    (c) $t = \frac{2s}{(u + v)}$ (d) $s = \frac{(v^2 - u^2)}{2a}$

5   (a) $n = \frac{P}{h} - 2$ (b) $x = \frac{1}{2} - \frac{t}{6}$

52. Using algebra
1   (a) $7x - 9 = 96$ (b) 15
2   15 m
3   6 cm
4   16 m and 15 m

53. Identities
1   Could be odd or even.
    For example, when $p = 2$, $q = 7$, $p^2 + 3q = 25$;
    when $p = 3$, $q = 7$, $p^2 + 3q = 30$
2   $h = 31$, $w = 5$
3   $x^2 + yz$
4   (a) $(x - 2)^2 = (x - 2)(x - 2) = x^2 - 2x - 2x + 4 = x^2 - 4x + 4$
    (b) $(x + 2)^2 - (x - 2)^2 = (x + 2)(x + 2) - (x - 2)(x - 2)$
       $= x^2 + 2x + 2x + 4 - (x^2 - 2x - 2x + 4)$
       $= x^2 + 4x + 4 - x^2 + 4x - 4$
       $= 8x$
5   $(y + 2)(y - 8)$

54. Problem-solving practice 1
1   8 days
2   (a) (6, 1)   (b) (4, 3)
3   (a) £130   (b) 6 hours
4   Dan is correct
    $4 \times 2 + 5 \times 2^2 = 8 + 5 \times 4 = 8 + 20 = 28$
5   AB = BC = 22.5 cm, AC = 15 cm

55. Problem-solving practice 2
6   1225
7   55°
8   (a) $y = 5x + k$ where k is any number
    (b) $y = 3x - 5$
9   $3n - 1 = 234$, giving $n = 78\frac{1}{3}$ which is not an integer
10   -9

RATIO & PROPORTION

56. Percentages
1   (a) 4.5   (b) 11
2   (a) 62.5%   (b) 20%
3   (a) £7.68   (b) £103.68
4   37.5%
5   252
6   (a) 30.8%

    (b) $\frac{30}{100} \times 140 = 42$ (French)

      $\frac{60}{100} \times 180 = 108$ (German)

      $\frac{150}{320} \times 100 = 46.9\% = 47\%$

57. Fractions, decimals and percentages
1   (a) $\frac{6}{25}$ (b) $\frac{16}{25}$ (c) $\frac{17}{20}$
2   (a) $\frac{3}{10}$, 61%, 0.62   (b) 0.32, 33%, $\frac{7}{20}$

    (c) 37%, 0.38, $\frac{2}{5}$
3   £575
4   58
5   250

58. Percentage change 1
1   1.08
2   (a) 74.88   (b) 131.04   (c) 84.48   (d) 275.88
3   £126.90
4   (a) 20%   (b) 30%   (c) 35%   (d) 10%
5   28%
6   £23 320

59. Percentage change 2
7   Kelly-air
8   Postland
9   Footworld

60. Ratio 1
1   (a) 3 : 2   (b) 27 : 8   (c) 7 : 8
2   (a) £20 : £30   (b) £100 : £250 : £400
3   cheese = 16 g, peppers = 24 g
4   Paul = 48 miles, Faye = 60 miles
5   (a) 1 part is 24, 24 × 3 = £72
    (b) (3 × 24) + (4 × 24) + (9 × 24) = £384

61. Ratio 2
6   (a) 28 g   (b) 40 g
7   £4400
8   48 ÷ 12 = 4
    Flour = 20, margarine = 16 and sugar = 12
    Flour = 1825 ÷ 20 = 91.25, margarine = 700 ÷ 16 = 43.75
    and sugar = 250 ÷ 12 = 20.83
    Therefore, maximum number of cakes = 20
9   £45.75

62. Metric units
1   (a) 4.5 cm   (b) 720 mm   (c) 3500 m
    (d) 5300 g   (e) 4300 ml   (f) 0.48 g
2   (a) 150 mm   (b) 2.8 cm   (c) 1.8 kg
    (d) 2.8 km
3   15 200
4   16
5   82
6   No, can only fit 19

63. Reverse percentages

1 £60
2 £33 250
3 £33 600
4 £491.96
5 Alison invested £1650 and Nav invested £1680. Nav invested more than Alison.

64. Growth and decay

1 £17 569.20
2 $n = 3$
3 1500×1.035^8
4 (a) £12 528.15   (b) £6332.78
5 (a) 6.5%   (b) £2815.71

65. Speed

1 8.9 m/s
2 $\frac{80}{1.75}$
3 4 hours 18 minutes
4 240 km
5 $35 \div 0.25 = 140$ km/h
  140 is greater than 130
6 Karen has the lower average speed.
7 100 m race

66. Density

1 0.875 g/cm³
2 147 g
3 $\frac{5000}{8.96}$
4 432 g
5 5666.4 g
6 Gavin is not correct, it is bronze.

67. Other compound measures

1 400 N/m²
2 18 750 N/m²
3 0.00625 m²
4 1322 N/m²
5 $12\,\text{cm} \times 100\,\text{cm} \times 100\,\text{cm} = 120\,000\,\text{cm}^3$
  $120\,000 \div 2 = 60\,000$
  $60\,000\,\text{cm}^3 = 60\,000\,\text{ml}$
  $60\,000 \div 250 = 240\,\text{s}$
  240 seconds is 4 minutes

68. Proportion

1 £2.80
2 £8.25
3 13
4 £16
5
Large	Medium
£ : g	£ : g
4.80 : 200	4.50 : 175
0.024 : 1	0.026 : 1

The large basket is better value for money.
6 20 days
7 8 days

69. Proportion and graphs

1 1170 Newtons
2 7.5 ohms
3 $x = \dfrac{1}{2y}$
4 (a) $P = 19.2$ (b) $Q = 3$
(c)
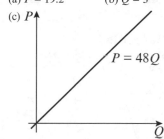
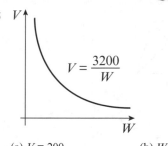

5
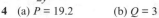

(a) $V = 200$ (b) $W = 2.5$

70. Problem-solving practice 1

1 Statistics
2 Nile is cheaper
3 46
4 $5400 \div 900 = 6$
  $\frac{1}{6}$ of $\frac{1}{5}$ of $900 = 30$ ml

71. Problem-solving practice 2

5 13 kg sack (£1.61 per kg; 5 kg sack is £1.80 per kg. Other methods available)
6 No. She needs $(60 \div 4) \times 15 = 225$ g of almonds and she only has 200 g.
7 No, he needs four more people.
8 (a) $\dfrac{100 + 6}{100} = 1.06$
  $£14\,000 \times (1.06)^4 = £17\,674.68$
  Kim has enough money.
  (b) 12 years (£28 170; 11 years is only £26 576)

GEOMETRY & MEASURES

72. Symmetry

1 (a)
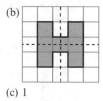
(b)
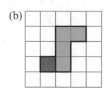

2 (a) 4   (b) 2 or 3   (c) 1
3 (a) 6
(b)
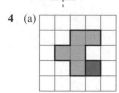
(There are other possible lines)

4 (a)
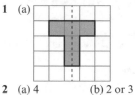
(b)

(c) 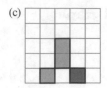 (d)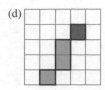

73. Quadrilaterals
1 (a) Rectangle   (b) Trapezium
   (c) Parallelogram   (d) Square
   (e) Rhombus   (f) Kite
2 (a) 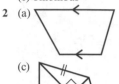   (b)

   (c)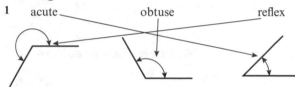

3 Rhombus

74. Angles 1
1 acute   obtuse   reflex

2 (a) Obtuse angle
   (b) $x°$ is more that 90° but less than 180°.
3 (a) Reflex angle
   (b) $x°$ is more than 180° but less than 360°.
4 (a) 109°
   (b) Angles on a straight line add up to 180°.
5 (a) 146°
   (b) Angles round a point add up to 360°.

75. Angles 2
6 292°
7 (a) 60°.
      The triangle is an equilateral triangle
   (b) 150°
8 (a) 65° because angles on a straight line add up to 180°.
   (b) 65° because x and y are alternate angles.
   (c) 65° because x and z are corresponding angles.

76. Solving angle problems
1 (a) 42°
   (b) alternate angles
   (c) 111 because angles on a straight line add up to 180°.
2 (a) 110°
   (b) corresponding angles
   (c) 40°
   (d) isosceles triangle
3 39°

77. Angles in polygons
1 (a) 72°   (b) 60°   (c) 45°
2 (a) 40°   (b) 9
3 (a) 12   (b) 10   (c) 20
4   360° ÷ 6 = 60° (exterior angle)
     180° − 60° = 120° (interior angle)
   180° − 120° = 60° (180° in a triangle)
        60° ÷ 2 = 30° (isosceles triangle)
5 (a) 15   (b) 24°

78. Time and timetables
1 (a) 15:15   (b) 02:25   (c) 23:48
2 (a) 4.25 am   (b) 12.10 pm   (c) 8.32 pm
3 52 minutes
4 15:30
5 (a) E   (b) 111 minutes
   (c) 07:45   (d) 09:32

79. Reading scales
1 (a) 23   (b) 340   (c) 5300   (d) 4.6
2 (a)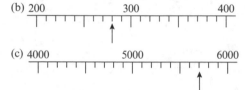

   (b)

   (c)

   (d)

3 (a) 65 km/h
   (b)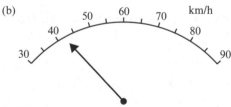

4 0.75 kg

80. Perimeter and area
1 (a) 5 cm²   (b) 12 cm
2 (a) 9 cm²   (b) 6 cm²
3 The perimeter of X is the same as the perimeter of Y.

81. Area formulae
1 (a) 90 cm²   (b) 36 cm²   (c) 28 cm²   (d) 190 cm²
2 $x = 3.6$ cm
3 $h = 5$ cm

82. Solving area problems
1 115 cm²
2 No, at least another £183 is needed.
3 $h = 13$ cm

83. 3D shapes

1 (a) Cube (b) Cuboid (c) Cylinder
   (d) Triangular prism (e) Pyramid (f) Sphere
2

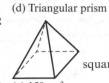

square-based pyramid

3 (a) 150 cm² (b) 122 cm² (c) 184 cm²
4 (a) 96 cm² (b) She does not have enough tins
              as 40 × 96 = 3840 cm²
              and 10 tins = 3500 cm²

84. Volumes of cuboids

1 (a) 125 cm³ (b) 360 cm³ (c) 1344 cm³
2 6 cm
3 378
4 11.67 cm

85. Prisms

1 (a) 48 cm³ (b) 144 cm³ (c) 384 cm³
2 (a) 108 cm² (b) 216 cm² (c) 366 cm²
3 592 cm³

86. Units of area and volume

1 (a) 60 000 cm² (b) 1500 mm² (c) 4 000 000 m²
   (d) 50 m² (e) 600 cm² (f) 0.8 km²
   (g) 2.75 cm²
2 (a) 22 000 000 cm³ (b) 28 000 mm³
   (c) 3 000 000 000 m³ (d) 200 m³ (e) 50 000 cm³
   (f) 200 litres (g) 8000 litres
3 No, the first multiplication should be 100 × 100 × 100. The
   correct answer is 25 000 litres.
4 (a) 45 litres (b) 96 000 litres

87. Translations

1 (a) $\binom{3}{4}$ (b) $\binom{-2}{-3}$ (c) $\binom{-5}{6}$
2 (a) and (b)

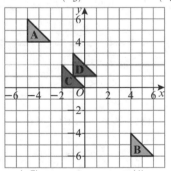

3 (a) translation $\binom{-7}{-3}$ (b) translation $\binom{4}{5}$
4 $\binom{-8}{-3}$

88. Reflections

1 (a)

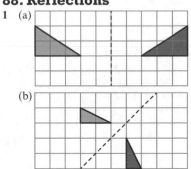

(b)

2 (a), (b) and (c)

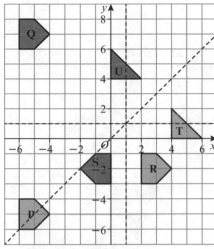

3 (a) Reflection in the line $y = -1$
   (b) Reflection in the line $y = x$
4

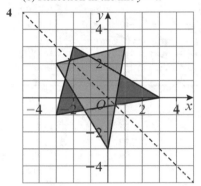

89. Rotations

1 (a) and (b)

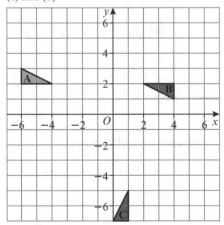

2 (a) and (b)

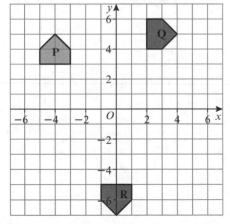

3 (a) Rotation 90° about (1, −1) clockwise
(b) Rotation 180° about (0, −1)

90. Enlargements

1 (a) 5
(b)

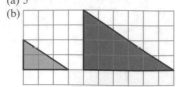

2

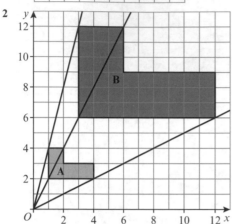

3 (a) Enlargement of scale factor $\frac{1}{3}$ at (4, −4)
(b)

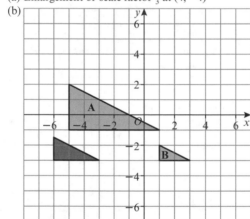

91. Pythagoras' theorem

1 (a) 10.7 cm   (b) 15.7 cm
2 She is incorrect as the length of the diagonal of her suitcase is 119.2 cm
3 No, $8.4^2 + 11.7^2 \neq 14.6^2$

92. Line segments

1 9.22
2 7.07
3 10
4 (a) (1, −2)
(b) diameter = $\sqrt{8^2 + 6^2}$ = 10
    hence, radius = 10 ÷ 2 = 5

93. Trigonometry 1

1 (a) 54.3°   (b) 57.8°
2 (a) Either comment that the ratio is upside down, or, remark that the bigger angle ought to be opposite the longer side and this is not true here.
(b) 54.1°
3 53.0°
4 She cannot use smooth tiles on her roof as angle $x°$ is greater than 17° (20.4°).

94. Trigonometry 2

5 (a) 14.3 cm   (b) 16.3 cm
6 5.7 m
7 7.3 m
8 (a) 22.4 m   (b) 61.8°

95. Solving trigonometry problems

1

	0°	30°	45°	60°	90°
sin	0	$\frac{1}{2}$	$\frac{1}{\sqrt{2}}$	$\frac{\sqrt{3}}{2}$	1
cos	1	$\frac{\sqrt{3}}{2}$	$\frac{1}{\sqrt{2}}$	$\frac{1}{2}$	0
tan	0	$\frac{\sqrt{3}}{3}$	1	$\sqrt{3}$	−

2 $x = 9$ cm
3 (a) 30°   (b) 30°
4 $30\sqrt{3}$
5 $\frac{44}{\sqrt{3}}$ m or $\frac{44\sqrt{3}}{3}$ m

96. Measuring and drawing angles

1 (a) 45°   (b) 140°   (c) 225°
2 (a)

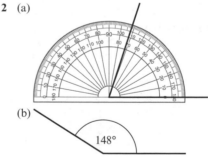

(b)

148°

3 66°
4 (a) 55°, acute   (b) 150°, obtuse   (c) 235°, reflex
5

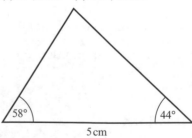

58°   44°
5 cm

97. Measuring lines

1 (a) 3.2 cm   (b) 4.3 cm   (c) 5.9 cm
2 (a) A line 52 mm long (±1 mm)
(b) A line 6 cm long (±1 mm)
(c) A line 7.8 cm long (±1 mm)
3 cross marked half way between A and B (±1 mm)
4 6.4 m
5 (a) 2 m   (b) 6 m

98. Plans and elevations

1

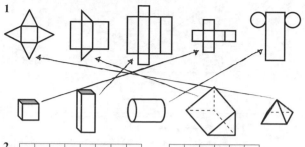

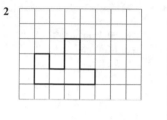

2

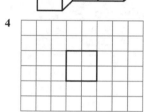

3 (a) 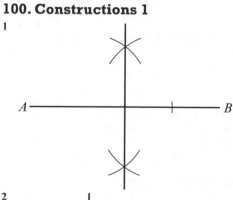 (b) 5

4

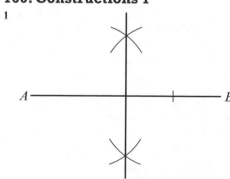

99. Scale drawings and maps

1	(a) 35 m	(b) 22.5 km	(c) 78 km
2	(a) 2.5 km	(b) 12 km	(c) 154 km
3	(a) 20 cm	(b) 15 cm	(c) 5 cm
4	230 mm		

100. Constructions 1

1

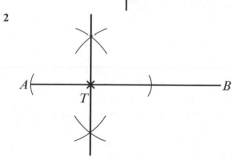

2

A $\left(\!\left.\times\atop T\right.\!\right)$ B

3

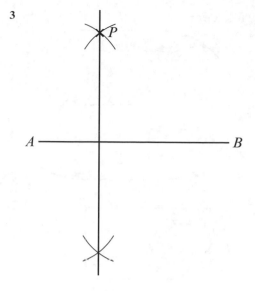

101. Constructions 2

4

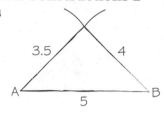

5

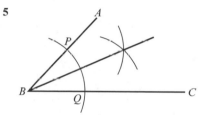

6

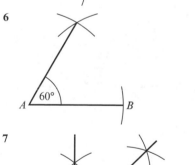

7

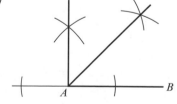

151

102. Loci

1

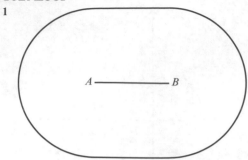

2

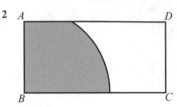

3

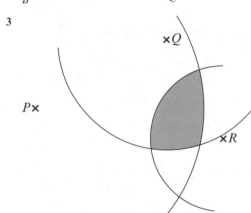

4

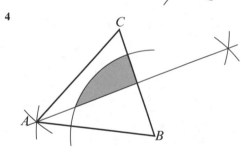

103. Bearings

1 (a) 270°   (b) 225°   (c) 135°
2 (a) 30°   (b) 210°
3 (a) 130°   (b) 310°
4 (a) N   (b)   N   (c)   N

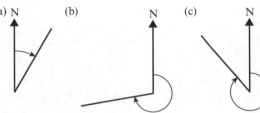

5 (a) 024°
(b)

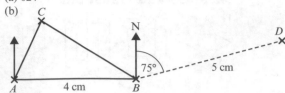

6 065°

104. Circles

1 (a) chord   (b) radius   (c) diameter
   (d) tangent
2 (a) 37.7 cm   (b) 50.3 cm
3 (a) 5.57 cm   (b) 14.6 cm
4 (a) 42.8 cm   (b) 72.0 cm
5 3140 cm

105. Area of a circle

1 (a) 113 cm²   (b) 201 cm²   (c) 531 cm²
2 (a) 113 cm²   (b) 308 cm²   (c) 1360 cm²
3 (a) 204 cm²   (b) 42.1 cm²
4 Both areas = 25π cm² or 78.53... cm²

106. Sectors of circles

1 (a) 5.59 cm   (b) 34.0 cm
2 (a) 24.4 cm   (b) 58.5 cm
3 (a) 36.3 cm²   (b) 145 cm²
4 $\dfrac{195}{360} \times 2 \times \times 8$
5 107.4°

107. Cylinders

1 (a) 704 cm³   (b) 11 300 cm³   (c) 25 400 cm³
2 (a) 452 cm²   (b) 2790 cm²   (c) 4810 cm²
3 (a) cylinder = $\pi \times 15^2 \times 18 = 12\,723$ cm³
       cube = $24 \times 24 \times 24 = 13\,824$ cm³
       Volume of the cube is greater.
   (b) cylinder = $(2 \times \pi \times 15 \times 18) + (2 \times \pi \times 15^2) = 3110.2$ cm²
       cuboid = $24 \times 24 \times 6 = 3456$ cm²
       The cuboid has the greatest surface area.

108. Volumes of 3D shapes

1 (a) 251 cm³   (b) 7240 cm³   (c) 142 cm³
2 (a) He uses $\frac{4}{3}$ instead of $\frac{2}{3}$
       He uses 6 instead of 3 for the radius.
   (b) 18π cm³
3 Volume of cylinder = $240\,\pi$ cm³
   Volume of cone = $120\,\pi$ cm³
4 300 cm³

109. Surface area

1 (a) 251 cm²   (b) 1810 cm²   (c) 462 cm²
2 320 cm²
3 905 cm²
4 96π cm²

110. Similarity and congruence

1 (a) A and B   (b) D and E

2 (a)

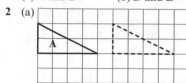

(b)

3 (a)

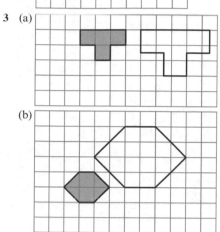

(b)

5 27

6 B

111. Similar shapes

1 (a) 130°   (b) 30 cm   (c) 18 cm

2 (a) 33 cm   (b) 8 cm

3 (a) 6.4 cm   (b) 5.7 cm

112. Congruent triangles

1 Side AC = side DF, side AB = DE, side BC = EF therefore SSS

2 BC = QR, CA = PR, angle BCA = angle PRQ therefore SAS

3 A and D

113. Vectors

1 (a) $\binom{2}{5}$ (b) $\binom{-2}{-5}$ (c) $\binom{5}{3}$

   (d) $\binom{-5}{-3}$ (e) $\binom{5}{-6}$ (f) $\binom{-5}{6}$

2 (a) $\binom{-20}{30}$ (b) $\binom{10}{2}$ (c) $\binom{-19}{20}$

3 (a) $\mathbf{p} + \mathbf{q}$ (b) $-\mathbf{q} - \mathbf{p}$ (c) $\mathbf{p} - \mathbf{q}$ (d) $\mathbf{q} - \mathbf{p}$

4 $\mathbf{b} - 3\mathbf{a}$

114. Problem-solving practice 1

1 18°

2 tray = 60 cm × 40 cm × 2 cm = 4800 cm³
cylinder = $\pi \times 9^2 \times 20$ = 5089 cm³
There will be no water left in the rectangular tray.

3 (a) and (b)

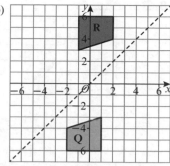

(c) Reflection in the line $y = x$

115. Problem-solving practice 2

4 Yes, she does have enough bags.

5 Large trapezium = $\frac{1}{2} \times 9 \times (24 + 18)$ = 189 cm²
Small trapezium = $\frac{1}{2} \times 6 \times (16 + 12)$ = 84 cm²
Difference in area = 105 cm²

6 (a) 62.0°
(b) $\sin 42° = \dfrac{12.8}{BD}$ therefore BD = 19.1 m
It is long enough.

PROBABILITY & STATISTICS

116. Two-way tables

1

	Bath	Warwick	Lichfield	Total
Boys	10	14	8	32
Girls	7	11	20	38
Total	17	25	28	70

2 (a)

	Dodgeball	Football	Rounders	Total
Girls	12	18	11	41
Boys	6	19	14	39
Total	18	37	25	80

(b) 19   (c) 41   (d) 12

3 (a)

	white	blue	red	Total
Motorbikes	7	9	6	22
Cars	3	8	17	28
Total	10	17	23	50

(b) 22   (c) 28   (d) 20

117. Pictograms

1 (a) 8 hours   (b) 3 hours

(c)

Monday	○ ○ ○ ○
Tuesday	○ ○ ○
Wednesday	○ ○ ◖
Thursday	○ ○
Friday	○ ◖

2 (a) 20 packets   (b) 25 packets

(c)

Monday	▭ ▭ ▭ ▭
Tuesday	▭ ▭ ▭
Wednesday	▭ ▭ ▪
Thursday	▭ ▭ ▭ ▭
Friday	▭ ▭ ▭

118. Bar charts

1 (a)

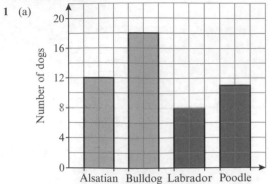

(b) Bulldog   (c) 49

2 (a) Dal   (b) 4

(c)

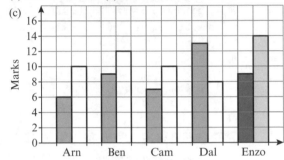

3   1. The scale on the *y*-axis is not linear.
     2. One of the bars is not labelled.

119. Pie charts

1

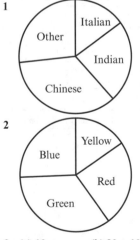

2

3 (a) 10   (b) 20 + 10 + 30 + 60 = 120

120. Scatter graphs

1 (a) Positive   (b) 135 g   (c) 225 g   (d) Yes
(e) The reading is not within the range of data and we are having to extend the line of best fit.

2 (a) Negative   (b) £1250
(c) This reading is reliable.   (d) No

121. Averages and range

1 (a) 7   (b) 11   (c) 10
2 (a) £336
(b) £756 is an outlier and has caused the mean to be high and not typical of the wages of the other five people. (Mean of the other five = £252)
3 (a) 6, 6, 9   (b) 7, 7, 8, 11, 12

122. Averages from tables 1

1 (a) 1   (b) 1   (c) 1.6   (d) 4
2 (a) 5   (b) 5   (c) 4.12
3 Mean mark = 385 ÷ 25 = 15.4, so 3 + 6 + 2 = 11 students have to re-take the test.

123. Averages from tables 2

4 (a) $8 \leqslant h < 10$ (b) $6 \leqslant h < 8$ (c) $\frac{195}{35} = 5.57$
(d) Because we are taking the midpoint.
5 (a) 21 minutes
(b) Yes because 32 minutes is greater than the mean.

124. Line graphs

1 (a)

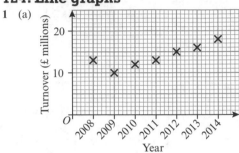

(b) Upwards
2 (a) 4   (b) 60
3 (a) 6   (b) 6.2

125. Sampling

1 (a) It is quick, cheap and easier to handle.
(b) 5.14
(c) Not very reliable as sample is small
(d) Ask children in different classes in different years.
2 (a) 21.5 cm
(b) 1.7 m
(c) Part (a) is more reliable as it is within the data range.
(d) Carry out more experiments

126. Comparing data

1 (a) Students did better in Maths because the mean was higher. Students' results in Maths were less varied because the range was smaller.
(b) The amount of rainfall was higher in Dundee because the mean was higher. Wolverhampton's amount of rainfall was more varied because the range was higher.
2 (a) 3.25   (b) 8
(c) Mr Jones's class had fewer absences, but these were spread over a wider range.
3 (a) Alex scored 380, Joe scored 362 (45.25 × 8), so Alex scored 18 runs more than Joe.
(b) Alex's range was 90 (94 − 4), Joe's range was 38; so Joe's scores vary less than Alex's scores, meaning he is the more consistent player.

127. Probability 1

1 (a)

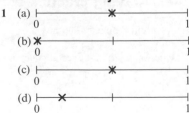

2 (a) Impossible   (b) evens   (c) certain

3 (a) $\frac{1}{6}$ (b) $\frac{1}{2}$ (c) $\frac{1}{2}$ (d) 0

4 (a) $\frac{4}{8}$ (b) $\frac{1}{8}$ (c) 0   (d) $\frac{7}{8}$

128. Probability 2

5 25%

6 (a) 0.21   (b) 0.74

7 0.08

8 $\frac{18}{96}$ or $\frac{3}{16}$

129. Relative frequency

1 (a) $\frac{1}{5}$ (b) $\frac{17}{50}$ (c) $\frac{41}{50}$

2 (a) $\frac{140}{200}$ or 0.7

    (b) The sample is large so the estimate is accurate.

3 (a) $\frac{8}{60} = \frac{2}{15}$ (b) $\frac{1}{6}$

    (c) Yes. If spinner is fair would expect to spin a 1 about half the time. (Relative frequency of spinning 1 = $\frac{34}{60}$)

    (d) 25

130. Frequency and outcomes

1 (C, P) (C, G) (C, B) (L, P) (L, G) (L, B) (V, P) (V, G) (V, B)

Probability = $\frac{1}{9}$

2 (a) $\frac{1}{3}$

    (b)

Neil's card	X	X	X	Y	Y	Y	Z	Z	Z
Tej's card	X	Y	Z	X	Y	Z	X	Y	Z

    (c) $\frac{1}{3}$ (d) $\frac{2}{3}$

3 (a)

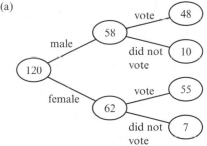

    (b) $\frac{10}{58} = \frac{5}{29}$

4 35

131. Venn diagrams

1 (a) (i) $x = 15$ (ii) $x = 9$ (iii) $x = 9$

    (b) (i) Students who only study maths

        (ii) Students who don't study French or German

        (iii) Students who study both DT and ICT

2 (a) $\frac{1}{8}$ (b) $\frac{11}{40}$ (c) $\frac{21}{40}$

3 (a)

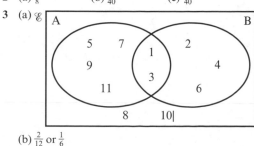

    (b) $\frac{2}{12}$ or $\frac{1}{6}$

132. Independent events

1 (a) $\frac{9}{100}$ (b) $\frac{49}{100}$ (c) $\frac{42}{100}$

2 (a)

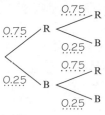

    (b) 0.625

3 (a)   Nav   Asha   (b) 0.26

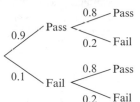

133. Problem-solving practice 1

1 (a) Box D, probability of $\frac{3}{5}$ is the greatest

    (b) Boxes A and C, probability of $\frac{1}{3}$

2

	French	German	Spanish	Total
Female	15	11	13	39
Male	16	17	8	41
Total	31	28	21	80

    (b) $\frac{31}{80}$

3 (a)

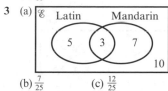

    (b) $\frac{7}{25}$ (c) $\frac{12}{25}$

134. Problem-solving practice 2

4

Favourite snack in year 11	Frequency	Angle
Burger	40	80°
Chips	90	180°
Hot dog	20	40°
Kebab	30	60°
Total	180	

5 (a) 0.75   (b) 0.2   (c) 30

6 (a)   Jamie   Rajiv

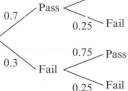

    (b) $\frac{2}{5}$

PRACTICE PAPER

Paper 1

1 (a) 14 km

(b) Harry; he walks 39 km , Lewis walks 37 km

2 $\frac{20}{25}$

3 $\frac{1}{4}$

4 64

5 (a) 7°C   (b) −11°C   (c) 6°C   (d) 10°C

6 (a) $7 \times (8 - 5) + 3 = 24$

(b) $(9 + 1) \times 8 \div (4 - 2) = 40$

7 260

8 (a) $10x$ (b) $20e - 8f + 4$

9 110°. Other two angles in isosceles triangle are each 70°, then use angles on a straight line.

10 5.5 cm = 5.5 × 10 000 = 55 000 cm = 550 m

11

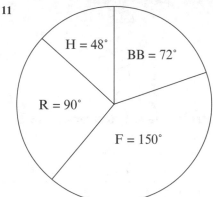

12 84

13 (a) $5(x + 2)$ (b) $x(x - 6)$

14 0.3

15 £519 to pay. He saves £42 per week.

Number of weeks = 519 ÷ 42 = 12.35 ...... = 13 weeks

16 (a)

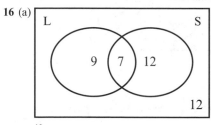

(b) $\frac{12}{40} \times 100 = 30\%$

17 90 km/hour

18 16

19 (a) $\dfrac{1\frac{1}{2}}{2\frac{1}{2}} = \frac{3}{5}$

(b) $900 \equiv \frac{2}{5}$; Total = 900 ÷ 2 × 5 = 2250

20 −3

21 £459

22 $4n + 3$

23 $\frac{1}{4} \times \pi \times 122 = 36\pi$ cm²

24 $\sum fx = 204$; Mean = 204 ÷ 30 = 6.8 hours

25 1   4   16   64

26 (a) 5×10^4 (b) 0.000096   (c) 1.5×10^{11}

27 (a) 120°   (b) 10 cm   (c) 27 cm

28 (a)

x	0.5	1	2	4	5	8
y	16	8	4	2	1.6	1

(b)

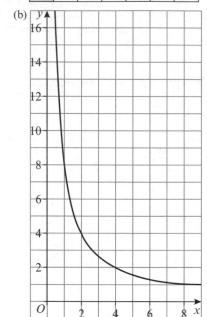

Published by Pearson Education Limited, 80 Strand, London, WC2R 0RL.

www.pearsonschoolsandfecolleges.co.uk

Text and illustrations © Pearson Education Limited 2017

Typeset and illustrated by Tech-Set Ltd, Gateshead

Produced by Out of House Publishing

Cover illustration by Miriam Sturdee

The right of Glyn Payne to be identified as author of this work has been asserted by him in accordance with the Copyright, Designs and Patents Act 1988.

Content written by Navtej Marwaha is included.

First published 2017

20 19 18 17
10 9 8 7 6 5 4 3 2 1

British Library Cataloguing in Publication Data
A catalogue record for this book is available from the British Library

ISBN 978 1 447 98786 4

Printed in Slovakia by Neografia